IEE Materials & Devices Series 4

Series Editors: Dr N. Parkman
            Professor D. V. Morgan

SEMICONDUCTOR LASERS FOR LONG-WAVELENGTH OPTICAL-FIBRE COMMUNICATIONS SYSTEMS

Other volumes in this series

SEMICONDUCTOR LASERS FOR LONG-WAVELENGTH OPTICAL-FIBRE COMMUNICATIONS SYSTEMS

M. J. Adams, A. G. Steventon,
W. J. Devlin and I. D. Henning

Peter Peregrinus Ltd. on behalf of The Institution of Electrical Engineers

Published by: Peter Peregrinus Ltd., London, United Kingdom

© 1987: Peter Peregrinus Ltd.

British Library Cataloguing in Publication Data

Semiconductor lasers for long-wavelength optical
    fibre communications systems.—(IEE
    materials and devices series; 4).
    1. Semiconductor lasers
    I. Adams, M.J.   II. Series
    621.36'61   TA 1700

ISBN 0 86341 109 6

Printed in England by Short Run Press Ltd., Exeter

Contents

Preface

The worldwide interest in optical-fibre communications has moved very rapidly from systems using a wavelength of about 0.85 μm to those using 1.3 or 1.55 μm, because of the greater repeaterless system lengths and/or bandwidths which they offer. The shorter-wavelength systems, which use silicon detectors and GaAlAs lasers or LEDs, have been in use in public systems since 1980 and will continue to be economically preferred for short links, i.e. from a few centimetres for communications on a single printed-circuit board to links of a few kilometres. The higher-performance systems required development of both emitters and detectors. Although the considerable experience on GaAlAs lasers laid the foundation for the development of longer-wavelength emitters, it had to be accomplished in the hitherto undeveloped material system of GaInAsP on InP substrates. For the highest-performance systems, considerable improvements in the spectral stability and purity were needed.

This book reviews the progress in the development of long-wavelength lasers for optical-communications systems. First we give a historical survey of the development of lasers and optical fibres; next we outline the long-wavelength systems possibilities and their demands on laser sources. We describe the important fundamental aspects of lasers in Chapter 3, followed by a description of the physical structures and fabrication techniques which have been used to realise practical lasers in Chapters 4 and 5. Chapter 6 very briefly summarises the international status of laser reliability, while Chapter 7 reviews the spectral and transient behaviour of long-wavelength lasers. Finally we consider some of the possible trends for the future and outline their impact on laser specifications.

Acknowledgments

The authors would like to express their gratitude to M. G. Burt, M. J. Robertson, A. W. Nelson and to many other colleagues at British Telecom Research Laboratories for invaluable discussions. Acknowledgment is made to the Director of Research of British Telecom for permission to publish.

Chapter 1

Historical survey

1.1 Semiconductor-laser development

It has been claimed that 'the pattern of independent multiple discoveries in science is . . . the dominant pattern rather than a subsidiary one' (Merton, 1961). The early history of semiconductor-laser development provides, as we shall see, at least three examples in support of this claim. Indeed, the first example is provided by the initial reports of lasing action in semiconductors by three groups working at IBM (Nathan et al., 1962), General Electric (Hall et al., 1962: Holonyak and Bevacqua, 1962) and MIT Lincoln Laboratory (Quist et al., 1962). The dates of receipt of the papers referred to were, respectively, 4 October, 24 September, 17 October and 23 October. However, it is important to stress that this bald statement about the first reports of semiconductor lasers hides the fact that the idea had been 'in the air' for some years before, and that much speculation on how such devices might be constructed had already been published (Basov et al., 1959, 1961; Bernard and Duraffourg, 1961; Watanabe and Nishizawa, 1957, 1960). It is not our purpose here to discuss the story of these early years in more depth, but the interested reader will find further details in the book by Casey and Panish (1978) or in an introduction by Hilsum (1969). Also a first-hand account has been given by Hall (1976).

The structure of these early lasers is of interest in view of the many developments and refinements which were to follow later. The basic device consisted of a rectangular parallelepiped of the III–V semiconductor GaAs incorporating a p–n junction which was formed by diffusion of a p-type dopant (usually Zn) into an n-type substrate. Two opposite faces of the device are polished to form partially reflecting mirrors so that a Fabry–Perot cavity is produced with its axis in the plane of the p–n junction. Under conditions of sufficiently strong forward bias (with a voltage nearly equal to the energy gap), stimulated emission of near-band-gap radiation occurs in the vicinity of the junction. Lasing action is then produced by selective amplification of one or more of the cavity modes. The current density required to achieve lasing in these early diodes was of order 10^5 A/cm^2, so that it was necessary to operate at liquid-nitrogen temperature (77 K). The struggle to achieve sufficiently low threshold currents to permit continuous wave (CW) operation at room temperature was to take another eight years.

Whilst the first diode lasers had cavities which were formed by polishing the facets, it rapidly became the norm to produce the mirrors by cleaving along a crystal plane (usually the {110} plane when the junction is in the {100} plane). An illustration of the structure of an early GaAs laser is given in Fig. 1.1. Another improvement in laser fabrication also occurred soon after the first observations of lasing action; this was the use of epitaxial growth to form the $p–n$ junction. Liquid-phase epitaxy (LPE) of GaAs was originally reported by Nelson (1963) and its subsequent use for laser fabrication enabled the threshold current density to be reduced to 3×10^4 A/cm^2 at room temperature (Dousmanis *et al.*, 1964). A 'melt' of gallium containing GaAs and the dopant at 900°C is brought into contact with a GaAs substrate and allowed to cool slowly. As the temperature falls, epitaxial growth of doped GaAs occurs and the laser junction is formed. More details of LPE and its application to long-wavelength laser fabrication are given in Section 5.1.

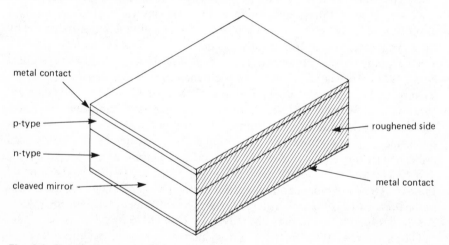

Fig. 1.1 *GaAs homostructure laser*

Although LPE growth resulted in lower thresholds, the goal of CW room-temperature operation remained elusive. However, another significant development occurred in 1967 when the use of a stripe-contact configuration resulted in CW operation of GaAs lasers at temperatures up to 205 K (Dyment and D'Asaro, 1967). The stripe-geometry homostructure laser is illustrated in Fig. 1.2. Its primary advantages lie in restricting the total device current and improving the heat-sinking properties of the laser. A secondary advantage is that this technique produces a well-defined region laterally where light emission can take place, and thus avoids 'filamentary' emission which commonly occurred in broad-contact lasers (see Section 3.7).

The real breakthrough in laser development came with the use of heterojunctions. The suggestion of using a 3-layer structure with a thin layer of lower band-gap material between two higher-gap semiconductors had already been made in 1963 (Kroemer, 1963; Alferov and Kazarinov, 1963), but it was not

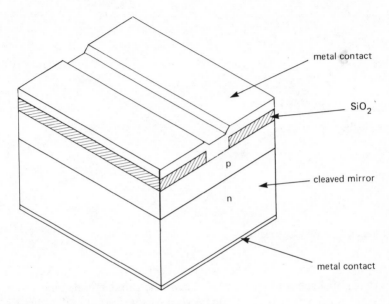

Fig. 1.2 *Stripe-geometry homostructure laser*

until the successful LPE growth of AlGaAs on GaAs (Rupprecht *et al.*, 1967; Alferov *et al.*, 1968) that this approach became technologically feasible. The use of heterojunctions was crucial for laser development in providing a means of confining the injected carriers (electrons in the conduction band and holes in the valence band) to a well-defined narrow layer where radiative recombination could occur. The lattice constants of AlAs and GaAs are sufficiently similar (less than 0.2% mismatch) that heterojunctions formed in this materials system exhibit good morphology and excellent electrical characteristics.

The fact that the energy-gap of $Al_xGa_{1-x}As$ increases monotonically with AlAs mole fraction x up to a value approximately 0.5 eV above that of GaAs (at which point the material becomes indirect gap) means that heterojunctions with extremely good carrier confinement can be made. The fraction of the total gap difference which is taken up by the conduction band as opposed to the valence band is still a matter of some controversy (see, e.g., Miller *et al.*, 1984) but it is certain that adequate barriers can be produced for both carrier types in appropriately designed structures.

The first diode laser to incorporate a heterojunction was the single-hetero-structure device shown in Fig. 1.3 (Kressel and Nelson, 1969; Hayashi *et al.*, 1969). Incidentally this device provides the second example of 'multiple independent discoveries' referred to in our opening sentence, since the two papers

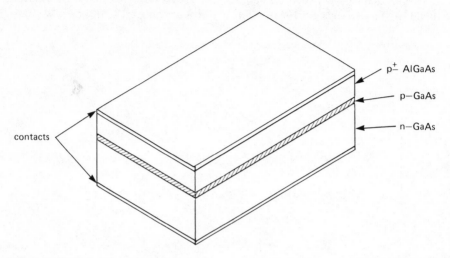

Fig. 1.3 *Single heterostructure laser*

announcing its successful operation independently at RCA and Bell Laboratories were published within a month of each other. In the single-heterostructure laser the heterojunction is positioned a short distance (optimally about $2\,\mu m$) away from a p–n junction in GaAs. Under forward bias, electrons injected across the p–n junction are prevented from travelling far in the p-region by the hetero-barrier. Lasing occurs in the narrow layer between the two junctions. Another effect of the structure arises from the fact that the refractive index of AlGaAs is lower than that of GaAs, and this leads to good optical confinement of the laser mode. In effect, a dielectric waveguide is formed in the structure, since the p-GaAs region is compensated and thus has a slightly higher refractive index than the n-region; the structure therefore consists of a high-index layer (p-GaAs) sandwiched between a layer of slightly lower index (n-GaAs) and one of considerably lower index (p-AlGaAs). This asymmetric dielectric waveguide acts to confine the optical intensity largely to the region where radiative recombination, i.e., gain, is occurring. A further advantage of the single-heterostructure over the homostructure device (Fig. 1.1) is that the optical loss suffered by the evanescent tail of the mode travelling in the AlGaAs region is somewhat smaller than that of the corresponding mode tail in the p-GaAs region of the homostructure, as a direct consequence of the difference in band gaps and hence of inter-band absorption.

Whilst the lowest room-temperature threshold current density obtained with the single-heterostructure device ($8600\,\text{A/cm}^2$) was considerably better than anything obtained with homostructure lasers, it was still too large to permit CW operation at room temperature. This goal was finally achieved with the double-heterostructure laser illustrated in Fig. 1.4 (Hayashi *et al.*, 1970; Alferov *et al.*, 1970), and yet again the result was achieved nearly simultaneously in two independent laboratories; in this case Bell Laboratories and the Ioffe Institute of Leningrad. In the double-heterostructure device the *p–n* homojunction is replaced by a second heterojunction between *n*-AlGaAs and *p*-GaAs. This has the effect of establishing symmetry for the confinement of both carrier types, since the progress of holes injected from the *p*-AlGaAs is now impeded by the second heterojunction. Similarly, the optical confinement is improved since the corresponding dielectric waveguide for the double heterostructure is symmetrical and the wave is more tightly confined to the active layer than was the case for the single heterostructure. The combined effects of improved carrier and optical confinement permitted room-temperature threshold current densities to be decreased to around $1600\,\text{A/cm}^2$, which was sufficiently low to permit CW lasing action when the device was mounted on a suitable heat sink.

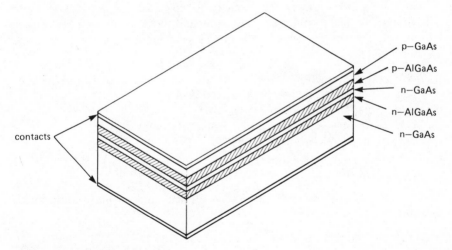

p–GaAs
p–AlGaAs
n–GaAs
n–AlGaAs
n–GaAs
contacts

Fig. 1.4 *Double heterostructure laser*

The achievement of CW operation gave a tremendous impetus to semiconductor laser research and enabled the main thrust of the work to be turned away from the preoccupation with threshold currents towards other topics, such as control of the lasing mode, achievement of high output power, improvement of the modulation characteristics, and studies of the laser spectral behaviour. As a consequence a diverse range of laser structures, of considerably more sophistication than the early ones, was developed. The process began with the adaptation of the stripe geometry, first used for homostructure devices, for incorporation in the new double heterostructures. Using a stripe contact formed by proton bombardment of the surrounding regions (rather than by oxide insulation as hitherto), the threshold current was reduced and a stable optical mode distribution produced in the lateral direction, i.e. along the heterojunction plane (Dyment *et al.*, 1972). The lateral optical confinement occurs by a process of 'gain-guiding' since there is optical gain in the region under the stripe, and loss elsewhere along the heterojunction plane. This results in a rather weak waveguiding effect, the strength of which varies with the operating current above threshold. As a result, the optical output power is not always linear with drive current, and sometimes the non-linearity can be especially marked, giving rise to 'kinks' in the light current plot. A second aspect of these non-linear light current plots arises from the lack of lateral current confinement; current spreading can occur outside the stripe region, thus lending a further complication to the device behaviour.

The search for means to avoid the problems of lateral optical and current non-confinement led to a multiplicity of laser designs, many of which are discussed, in their long-wavelength versions, in Chapter 4. In the present introductory Chapter we will merely indicate the initial achievements in GaAs/AlGaAs lasers. Briefly, the structures may be separated into two classes: those where lateral optical confinement is the aim, and those where both optical and current confinement is provided. The first class includes cases where the stripe geometry is defined by

(i)   A lateral p–n junction (the so-called transverse-junction-strip (TJS) laser first demonstrated by Namizaki *et al.* (1974))
(ii) Oxygen implantation (Blum *et al.*, 1975)
(iii) Use of a rib-waveguide structure (Lee *et al.*, 1975)
(iv) Zinc diffusion (Yonezu *et al.*, 1977)
(v)   Growth of planar hetero-layers over a channeled GaAs substrate (the 'channeled-substrate-planar' (CSP) laser (Aiki *et al.*, 1977).

The second class of devices, those with optical and current confinement, includes:

(i)   The mesa-stripe laser (Tsukada *et al.*, 1972)
(ii) The buried heterostructure (BH) laser (Tsukada, 1974), with its many derivatives (see Chapter 4).

A comprehensive discussion of GaAs/AlGaAs stripe-geometry lasers will be found in the book by Thompson (1980); further discussion of the semiconductor-laser family tree will not be attempted here. Instead we will turn our attention to the impact of optical-fibre development on laser materials and designs.

1.2 Optical-fibre development

At about the same time in the early 1960s when the first successful operation of
GaAs lasers occurred, the idea of using light as the carrier in communications
systems was being tentatively assessed. The initial studies in this direction
concentrated on protecting the light beam from atmospheric effects by means of
a hollow pipe, either using a reflective lining to reduce losses (Eaglesfield, 1962),
or incorporating a periodic sequence of lenses to prevent beam spreading
(Goubau and Schwering, 1961). However, the problems of multipath dispersion
for the former case, and the degree of precision and stability necessary for the
second, seemed to present insuperable obstacles. In spite of this, the attractions
of optical communications in terms of high bandwidth and potential cost
savings were such that the search for a suitable transmission medium continued.
A significant advance occurred in 1966 when it was realised that an optical fibre
made from silica might potentially have sufficiently low loss to permit long-
distance transmission (Kao and Hockham, 1966). Hitherto such fibres had been
used in the form of bundles for medical endoscopes, but their high attenuation
at optical wavelengths made transmission impossible over distances more than
a few metres. Kao and Hockham pointed out that pure silica was potentially a
low-loss material, and estimated that if an attenuation of 20 dB/km could be
achieved, then glass fibres would become extremely attractive for long-distance
transmission. In addition to the bandwidth advantage already referred to, fibres
also offer additional bonuses in terms of size, flexibility and tolerance to
environmental conditions.

The Kao and Hockham (1966) paper stimulated considerable interest (particularly in the UK) in the potential of optical fibres for communication. This
interest was further enhanced by the report in 1970 from Corning Glass Works
of an optical fibre with a measured loss of 20 dB/km (Kapron *et al.*, 1970). The
detailed history of these early years of fibre development need not concern us
here, but first-hand accounts have recently been published (Dyott, 1986; Karbowiak, 1986; Gambling, 1986), and the interested reader will find more
information in these fascinating reminiscences. In many ways the year 1970
marked a turning point in the development of optical communications, since it
was then that the two important breakthroughs of CW room-temperature laser
operation (see Section 1.1 above) and of low-loss fibre fabrication occurred.
Thus a convenient source and a suitable transmission medium for communications systems had been demonstrated. The remaining necessary component –
a fast, sensitive photodetector – already existed in the form of the silicon *p–i–n*
photodiode, whose maximum responsivity with respect to wavelength was
conveniently (and fortuitously) matched to the GaAs emission wavelength of
$0.85 \mu m$.

Low-loss fibres consist of a circularly cylindrical core surrounded by a cladding of slightly (by at most a few per cent) lower refractive index. Thus the
structure forms a weak dielectric waveguide of circular cross-section. For core

diameters of about 50–60 µm the guide will propagate many hundreds, possibly thousands, of modes, whilst single-mode fibres can be made by making the core only a few microns in diameter, the precise value depending on the refractive index difference of core and cladding, and on the wavelength of operation. The low attenuations achieved in such fibres depend critically on achieving purity of the materials used for core and cladding; usually these are composed of silica with the addition of various dopants to increase or decrease the refractive index as required. The most widely used method of fabrication is by modified chemical-vapour deposition (MCVD), a process involving the deposition of layers of material on the inside of a hollow silica tube (sometimes on the outside of a carbon mandrel) followed by collapse of the tube at elevated temperature to form a solid rod or 'preform' from which the fibre may be drawn. Achievement of low loss is dependent on the control of the deposition process and on the purity of the starting materials used for the core, and also (especially for single-mode fibres where significant amounts of power propagate outside the core) for the cladding.

In the early 1970s fibre loss at 0.85 µm continued to fall as a consequence of the development of the MCVD process, until values of a few dB/km were routinely achieved at this wavelength. However, these values are close to the fundamental limit set by Rayleigh scattering (see Section 2.1), and to achieve further reduction it is necessary to go to longer wavelengths. Thus it was shown that fibre losses could be reduced to 0.5 dB/km at 1.27 µm (Kawachi *et al.*, 1977) and to 0.2 dB/km at 1.55 µm (Miya *et al.*, 1979). In fact, 1.55 µm is the wavelength of minimum attenuation for silica fibres, since at longer wavelengths absorption (caused by the tail of infra-red molecular vibrations) becomes significant.

A second factor influencing the operating wavelength of an optical fibre for long-distance transmission arises from material dispersion. Different wavelengths propagate with different velocities, and in a communications system the pulse spreading resulting for a non-monochromatic source can place a limit on the bit rate achievable in practice. Since many semiconductor lasers (especially the earlier designs) are far from monochromatic, the material dispersion is an important factor in determining the maximum bit-rate–distance product. If one plots time of flight versus wavelength for a silica fibre, as in Fig. 1.5, then it is found that the curve has a minimum at 1.27 µm. Thus a source with non-zero linewidth centred at this wavelength would transmit signals which would suffer negligible pulse spreading (Payne and Gambling, 1975).

From the above discussion we see that there are two wavelengths of especial importance for optical communications, both determined by fundamental properties of silica fibre:

(i) The wavelength of minimum attenuation (1.55 µm),
(ii) The wavelength of zero material dispersion (1.27 µm).

When the significance of these two wavelengths was discovered in the mid-1970s, GaAs/AlGaAs laser development was at its height, and the realisation that these devices would no longer fulfil the main requirements as sources for transmission systems came as an unwelcome surprise to many would-be laser manufacturers. Whilst other semiconductors had been investigated for laser action (and in fact lasing had been observed at 1.6 μm in InAsP as early as 1964 (Alexander *et al.*, 1964), there had previously been very little incentive to consider the wavelengths 1.3–1.6 μm, and very little was known about the properties of materials which might be used to make lasers at these wavelengths.

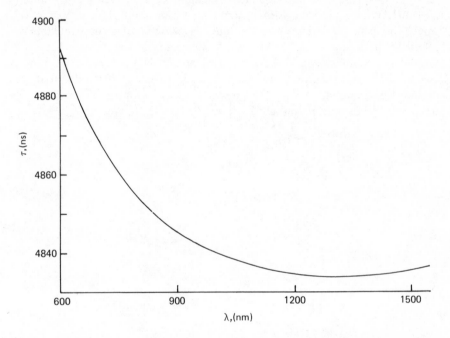

Fig. 1.5 *Time delay versus wavelength for a short pulse propagating in a 1 km length of optical fibre*

In fact, the choice of laser material is dictated not only by the emission wavelength but also by the requirement (see Section 3.3) to grow lattice-matched heterostructures on commercially available substrates. For the wavelengths of interest, the latter condition narrows the range of suitable III–V materials to only three: AlGaAsSb/GaAsSb grown on GaAs substrates (Nahory *et al.*, 1976). InGaAs/InGaP grown on GaAs substrates (Nuese and Olsen, 1975), and InGaAsP/InP grown on InP substrates (Hsieh *et al.*, 1976). Whilst CW room-temperature lasing has been determined in all three systems, the first two share a common problem of imperfect lattice match to the GaAs substrate. As a result, the material system which has received the most attention is InGaAsP/InP, where the first reports of CW room-temperature operation at 1.25 μm and 1.5 μm were published in 1977 (Hsieh and Shen, 1977) and 1979 (Kobayashi and Horikoshi, 1979), respectively.

Many of the laser designs already developed for the GaAs/AlGaAs system could be carried over directly to the long-wavelength material. An exception occurred for 1.5 μm lasers where it was found necessary to incorporate an extra layer of quaternary material to prevent meltback of the active layer during LPE growth. Two other significant differences between GaAs/AlGaAs and InGaAsP/InP lasers soon became apparent. The first was the enhanced sensitivity to temperature exhibited by long-wavelength lasers, a factor which implies a requirement for good heat-sinking and temperature control for many applications. The second difference was that long-wavelength lasers show much better reliability behaviour than GaAs/AlGaAs devices, and this feature gave considerable encouragement to the rapid development of long-wavelength lasers (see Chapter 6). In the remainder of this book we shall look in detail at that development, and summarise the present status of semiconductor lasers for long-wavelength optical communication.

Chapter 2

Optical-fibre systems

2.1 Fibre loss and dispersion

Optical-fibre communication systems use a fibre, an emitter and a detector. The fibre consists of a central core of very low-loss glass surrounded by a cladding layer having a lower refractive index, providing a low-loss optical waveguide. Modulated light from a laser or LED is coupled into one end of the fibre and detected at the other end. The fibre transmission losses due to absorption, scattering and radiation can be minimised by the use of high-purity materials, avoidance of inhomogeneities and the use of a sufficiently large refractive-index step between the core and cladding to avoid radiation, particularly at bends. The residual losses in modern silica-based fibres, illustrated in Fig. 2.1, are due to Rayleigh scattering from the fundamental inhomogeneities in the glassy matrix itself (which falls as λ^{-4}), the tail of the infra-red absorption band and impurity absorption due to residual hydroxyl ions. The peak at 1.1 μm is due to the cut-off of the LP_{11} mode. The resulting window has low losses from between about 1 and 1.7 μm with a minimum of just below 0.2 dB/km at 1.55 μm. The use of heavier glass matrices, which would shift the IR absorption tail to longer wavelength resulting in even lower minimum losses, will be discussed later (Chapter 8).

Dispersion is the other dominating fibre parameter. Fig. 2.2a illustrates a large core step-index fibre in which many guided modes can propagate, as indicated by the ray trajectories. Clearly, different modes will have different propagation constants resulting in modal dispersion, causing a temporal broadening of any input pulse of light regardless of its spectral purity. This dispersion can be dramatically reduced by using an approximately parabolic refractive-index profile as illustrated in Fig. 2.2b. Light travelling near the centre does so at a lower velocity than in the outer regions of the core, reducing dispersion by about 3 orders of magnitude. However the dispersion is very sensitive to the profile, and in practice fibre bandwidths are limited to about 4 GHz km, i.e. 0.25 ns broadening per km. The model dispersion can be eliminated

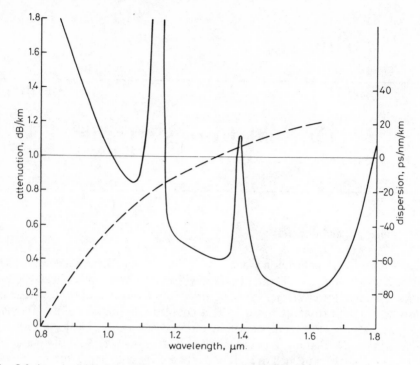

Fig. 2.1 *Loss (solid line) and dispersion (dashed line) of monomode silica fibre*

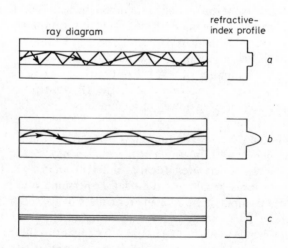

Fig. 2.2 *Ray diagram and refractive-index profile of (a) step-index multimode fibre, (b) graded-index multimode fibre and (c) step-index monomode fibre*

by reducing the core diameter until only one mode is guided (at a di
5–10 μm). This is the so-called monomode fibre.

The refractive indices of the fibre materials themselves are wavelength-depen-
dent, giving rise to the material dispersion which is plotted in Fig. 2.1 for silica.
It will be noted that the material dispersion passes through zero at approximately
1.3 μm. Material dispersion becomes a significant problem because conventional
Fabry–Perot lasers have a spectrum covering several Fabry–Perot modes. Operation at the zero-dispersion wavelength around 1.3 μm alleviates this problem.

In monomode fibres the evanescent tails of the optical wave penetrate the
cladding layer, which modifies the net dispersion since it has a different material
dispersion curve. This, together with the variation of the distribution of power
between the core and cladding with wavelength, gives rise to a waveguide
dispersion which increases with wavelength. The net effect is to increase the
wavelength at which the overall dispersion is zero. By manipulating fibre parameters the minimum in dispersion can be shifted to the minimum loss
wavelength of 1.55 μm. This could result in the ideal low-loss, low-dispersion
fibre; however, in practice such dispersion-shifted fibre has higher losses at
1.55 μm than standard fibre at 1.55 μm (Ainslie *et al.*, 1981; Tomaru *et al.*, 1981).
By grading the refractive-index profile, Ainslie *et al.* (1982) were able to achieve
low loss and low dispersion simultaneously at 1.55 μm. Other more complex
profiles in which dispersion is maintained at a low value over a wide wavelength
range are attractive for wavelength-multiplexed systems. Losses of 0.4 dB/km at
1.55 μm (Cohen *et al.*, 1982) and 0.24 dB/cm at 1.60 μm (Berkey, 1984) have
been obtained for such fibres, but they are much more difficult to produce and
control than step or triangular profile fibres.

2.2 Optical-fibre system limitations

The optical fibre system will be limited either by the optical power available or
by dispersion. The power available is usually expressed as the ratio of the power
launched into the fibre P_L and the minimum power required by the receiver to
maintain a bit-error ratio (BER) of 10^{-9}, P_R, both being expressed in dBm. This
'power budget' can be used to offset the fibre losses α_F dB/km due to the fibre
itself and any associated joints, and to provide a margin M to accommodate
component variations with time, temperature etc., to allow for repairs requiring
additional joints and extra fibre etc. Thus a system will be limited by power to
a length given by

$$L_{max,P} = \frac{P_L - P_R - M}{\alpha_F} \qquad (2.1)$$

The dispersion limitation will occur when the pulse spreading causes adjacent
pulses to overlap so that errors result (inter-symbol interference). For simplicity
we assume this occurs when a pulse broadens by half its initial width. For

monomode systems the net material/waveguide-pulse-broadening $\Delta\tau$ increases linearly as given by

$$\Delta\tau = \sigma\Delta\lambda L \tag{2.2}$$

where σ is the dispersion plotted in Fig. 2.1 in ps per km of fibre per nm of transmitter linewidth, $\Delta\lambda$ is the transmitter linewidth (nm) and L is the system length (km). For a system operating at a bit rate B per second, the dispersion-limited length would be

$$L_{max,D} = \frac{1}{2B\sigma\Delta\lambda} \tag{2.3}$$

These limits are schematically illustrated in Fig. 2.3.

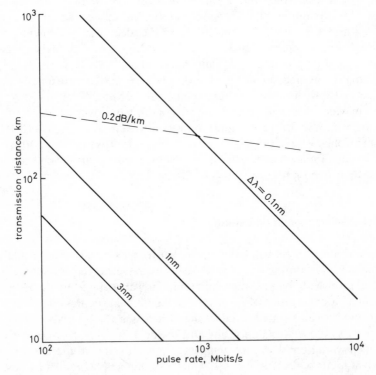

Fig. 2.3 *Example of loss (dashed lines) and dispersion (solid lines) limitations for silica monomode fibre at 1.55 μm*

The power limit can be improved by increasing the power emitted by the laser, improving the laser/fibre coupling efficiency, improving the receiver sensitivity, and by reducing fibre losses. The dispersion limit can be improved by reducing the source linewidth or/and operating at the zero dispersion point.

System parameters are combined in Table 2.1 to illustrate a range of possible systems. For simplicity $M = 0$, though a value of $M = 10\,\text{dB}$ may be required in some applications. The power limit (Fig. 2.3) reduces with increasing bit rate because the receiver sensitivity decreases by 3–4.5 dB per doubling of bit rate. In practice, the launched power may also decrease because of the increasing difficulty in modulating the laser. The Table shows clearly the system benefits realised by using longer wavelength, and also narrower laser linewidths.

The spectrum of a conventional Fabry–Perot laser contains power in several longitudinal modes, and when the laser is modulated directly the spectrum broadens. Eqn. 2.3 relies on the envelope of the time-averaged laser linewidth; however, it has been found experimentally that this is over-optimistic (see Chapter 7). This dispersion limitation is circumvented by using single-longitudinal-mode lasers (SLM lasers).

Even with SLM lasers dispersion results in a limitation. Under direct modulation the variation in current causes a variation of carrier concentration in the laser, resulting in a change in the refractive index and hence emission wavelength, called 'chirp' (Wright and Nelson, 1977). This chirp becomes significant only at bit-rate/length products of about 150–200 Gbit km. The actual amount of chirp depends on the laser structure and its operating conditions. Chirp, as well as transient intensity fluctuations, can be avoided by external modulation.

Early optical systems were based on multimode fibres because of their facility for low-loss fibre joints and ease of coupling to light sources. However, the dispersion analysis of these are more complex than indicated above (see e.g. Adams, 1981). For multimode systems, modal dispersion is usually dominant, but because power can be exchanged between fibre modes, particularly at bends or inhomogeneities, the pulse broadening does not increase linearly. In practice it increases with length to a power between 0·5 and 1. These factors do not of themselves impose any additional constraints on the laser specification, and hence are not dealt with further here.

The main emphasis of optical communication has been for digital modulation of the intensity of the light and direct intensity detection. Analogue modulation of short-wavelength lasers has received some attention and imposes considerable constraints on laser light/current linearity. Additionally, when there is some spatial filtering of the modes of a multimode fibre, e.g. at a joint, or coupling, modal noise (Epworth, 1978) is introduced because of the fluctuation of power in the various fibre modes. This modal noise gives rise to severe problems in analogue systems, which can be alleviated by using heavily multi-moded lasers, or SLM lasers and monomode fibres.

The direct intensity-detection technique ignores the wave nature of light. The phase and frequency information can increase the sensitivity of detectors by using homodyne or heterodyne principles. The resulting coherent fibre systems impose considerable constraints on SLM lasers.

Table 2.1 Possible System Budgets

λ (µm)	B (Mbits/s)	α_F (dB/km)	P_L (dBm)	P_R (dBm)	σ (ps/km/nm)	$\Delta\lambda$ (nm)	$L_{max.p}$ (km)	$L_{max.d}$ (km)	System
0.85	34	2.0	−10	−56	90	2	23	80	Early GaAs systems
0.85	140	2.0	−10	−50	90	2	20	20	
0.85	565	2.0	−10	−44	90	3	17	3.3	
1.3	140	0.4	0	−46	<4	3	115	146	'Zero' dispersion
1.55	140	0.25	0	−46	17	3	184	34	Minimum loss
1.55	140	0.25	0	−46	17	<0.1	184	>1000	Minimum loss and SLM laser
1.55	140	0.25	0	−62	17	<<0.1	248	>>1000	Minimum loss, coherent detection

2.3 Coherent optical-fibre systems

Homodyne or heterodyne detection offers improved receiver sensitivity because the square-law detection process (i.e. intensity rather than field) results in receiver gain when the detected signal is the product of the transmitted and a local-oscillator signal fields. The local oscillator is another laser, and its power, and hence the receiver gain, can be beneficially increased until its noise begins to swamp that of the detector and the following electronics. Improvements in receiver sensitivity of 10–17 dB over direct detection have been demonstrated (e.g. see review by Hooper *et al.*, 1983). These are within 4 dB of the quantum limit. The local-oscillator noise can be removed using a balanced receiver. Coherent techniques offer not only the benefit of increased system length (Table 2.1) but also the possibility of extensive multiplexing (there are about 70 000 1 GHz channels in the low-loss fibre window) and simple optical-amplifier repeaters. However, coherent systems impose considerable constraints on the components.

In the detector the local-oscillator and signal optical fields must be added with the same polarisation. In practice, polarisation fluctuations in standard fibres are relatively slow, and can be tracked with simple mechanical polarisation controllers (Harmon, 1982) although electro-optic ones are more likely to be engineered for operational use. The signal may be imposed on the carrier amplitude, frequency or phase resulting in amplitude, frequency or phase-shift key systems (ASK, FSK or PSK, respectively). Additionally the receiver could be either homodyne, where the local oscillator and signal oscillator have the same frequency, or heterodyne, where the frequencies differ. The most sensitive receiver is for PSK homodyne detection. All the coherent transmission options require SLM lasers, but additionally the linewidth and its stability must be controlled to be less than that of the modulation bandwidth, i.e. a control of much better than, say, 1 GHz in 200 THz; in fact, control of less than 100 kHz is required for PSK systems unless very rapid optical phase-locked loops can be developed. For ASK a linewidth of a few MHz is acceptable. Such control has in fact been achieved for semiconductor lasers and will be described in Chapter 7.

The chirp phenomenon mentioned earlier provides a convenient method of obtaining frequency modulation for FSK systems. Direct, but small, current modulation gives rise to a frequency change with a tolerably small amplitude change. In order to achieve ASK or PSK, separate modulators are required.

Stimulated Brillouin or Raman scattering are optically non-linear phenomena which impose limitations on the power density of spectrally pure light which can be transmitted through a fibre without suffering additional losses. Cotter (1983) showed that the stimulated Brillouin-scattering threshold power in a monomode fibre at 1.32 μm is only 5 mW. He also suggested ways of circumventing this. Stimulated Raman scattering would then impose an input power limit of about 500 mW (Stolen, 1980). These phenomena are important in coherent systems because the narrow linewidth necessary for them couples efficiently to the non-linear mechanism, whereas the broader linewidth usually used for direct detection does not. The use of an externally modulated narrow-line laser in direct detection systems could also suffer power limitation.

Chapter 3

Semiconductor-laser fundamentals

3.1 Radiative and non-radiative recombination

Efficient electroluminescence and low-threshold lasing action in semiconductors requires the dominance of radiative over non-radiative recombination. The principal recombination processes to be considered are illustrated in Fig. 3.1 for transitions between an upper energy level E_I and a lower level E_J. The upper set of states is assumed to be in quasi-equilibrium, characterised by a distribution function f_i and a quasi-Fermi level F_i; similarly the lower set is characterised by a distribution function f_j and a quasi-Fermi level F_j. In practice, the upper set of states is usually associated with the conduction band and the lower set is usually associated with the valence band, although shallow impurity levels (or impurity bands in the high-density case) are sometimes involved.

The process shown in Fig. 3.1a is made up of absorption of a photon of energy E ($= E_I - E_J$, by energy conservation), and the inverse process, induced emission of a photon of the same energy. The term 'stimulated emission' is reserved for the net downward transition rate, i.e. induced emission minus absorption. The probability of induced emission is found by multiplying the probability f_i of finding an electron in the upper level E_I by the probability $(1 - f_j)$ of finding no electron in the lower level E_J. Following a similar procedure for absorption, multiplying by the quantum-mechanical transition probability, and summing over all states I, J such that $E_I - E_J = E$, we arrive at the stimulated emission rate for photons of energy E, $r_{st}(E)$, as:

$$r_{st}(E) \propto f_i(1 - f_j) - f_j(1 - f_i) = f_i - f_j \qquad (3.1)$$

Hence, for a positive stimulated emission rate, we require $f_i > f_j$. Substituting the usual expressions for the Fermi–Dirac functions, this reduces to (Bernard and Duraffourg, 1961):

$$F_i - F_j > E \qquad (3.2)$$

This is a necessary, but not sufficient, condition for laser action in semiconductors. In order to achieve lasing, the stimulated emission rate must be sufficient in magnitude to overcome various loss mechanisms in the device, as we shall see in later subsections.

The process of Fig. 3.1b is that of spontaneous emission of a photon of energy E, and proceeds at a rate $r_{sp}(E)$, proportional to $f_i(1 - f_j)$. Since the constants

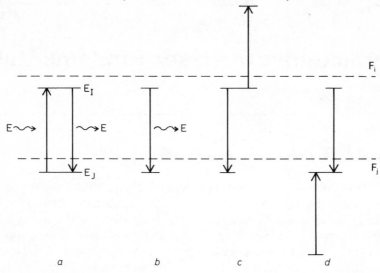

Fig. 3.1 *Recombination processes*
a absorption and induced emission
b spontaneous emission
c Auger recombination
d Auger recombination

of proportionality are the same for $r_{st}(E)$ and $r_{sp}(E)$ it follows that their ratio is given by (Lasher and Stern, 1964):

$$\frac{r_{st}(E)}{r_{sp}(E)} = \frac{f_i - f_j}{f_i(1 - f_j)} = 1 - \exp\left(\frac{E - (F_i - F_j)}{KT}\right) \tag{3.3}$$

where K is Boltzmann's constant and T is the absolute temperature. If the spontaneous rate $r_{sp}(E)$ is summed over all photon energies E, then we arrive at the total spontaneous emission rate R_{sp}. It is fairly easy to see intuitively (and in certain cases can be proved rigorously) that this will be proportional to the product of the electron concentration n for the upper levels with the hole concentration p for the lower levels:

$$R_{sp} = Bnp \tag{3.4}$$

where B is a constant of order 10^{-10} cm^3 s^{-1} for the semiconductors GaAs and InGaAsP at room temperature.

Turning now to non-radiative recombination, Figs. 3.1c and d illustrate two versions of the Auger effect (Beattie and Landsberg, 1958), the former involving collision of two electrons in the conduction band, whilst the latter involves the collision of a conduction-band electron with a low-energy valence-band electron. Once again we invoke an intuitive approach and argue that the process of Fig. 3.1c will have a total recombination rate R_{aug1} proportional to the product $n^2 p$, whilst for Fig. 3.1d the rate R_{aug2} will be proportional to np^2:

$$R_{aug1} = A_1 n^2 p; \qquad R_{aug2} = A_2 np^2 \qquad (3.5)$$

Other non-radiative recombination processes, such as those involving deep levels, multi-phonon emission, phonon cascade etc, are not considered here, and the interested reader is referred to a recent comprehensive review (Stoneham, 1981).

If one considers the quantum mechanics of the radiative and non-radiative transitions indicated in Fig. 3.1, then it can be shown that each process must conserve both energy and wave vector (or quasi-momentum) k. The implication of this for the radiative processes is illustrated in Fig. 3.2 for direct transitions between parabolic conduction and valence bands. On the scale of the electron and hole wave numbers ($\sim 10^9$ m^{-1}), the photon wave number is at least two orders of magnitude lower for visible or near infra-red radiation. Hence on

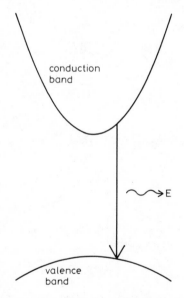

Fig. 3.2 *Radiative transition between parabolic conduction and valence bands*

the conventional E–k band diagram of Fig. 3.2 the photon wave number is neglected and the radiative transition is shown as a vertical line. Thus k-conservation immediately implies a requirement for direct-gap semiconductors if we wish radiative recombination to be significant. For an indirect semiconductor, where the valence-band maximum occurs at a different k from the conduction-band minimum, the excess wave vector in a radiative transition must be accounted for by the participation of a phonon. Since phonon-assisted radiative transitions are second-order processes from a quantum-mechanical viewpoint, they occur with much lower probability then direct recombination and are thus not suitable candidates as the basis of lasing action (Dumke, 1962).

For the Auger processes the analogous energy and wave-vector conservation rules lead to restricted sets of initial and final electron states for the various transitions, as illustrated in Fig. 3.3. In order to evaluate the most likely transitions it is necessary to take account of heavy (H) hole, light (L) hole and spin-orbit (S) split off valence bands as shown in the Figure. Fig. 3.3a illustrates the CHCC process which corresponds to the transition of Fig. 3.1c; only the conduction and heavy-hole bands are involved. By contrast the transition of Fig. 3.3b (CHSH) involves also the spin split-off band and is therefore

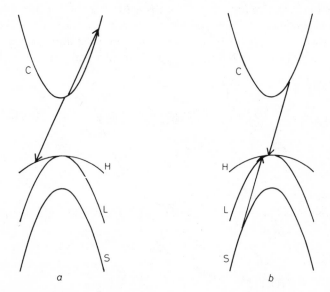

Fig. 3.3 *Auger transitions between conduction band (C), heavy hole (H), light hole (L), and spin split-off (S) valence bands*
a CHCC process
b CHSH process

likely to be significant for semiconductors where the energy gap E_G is not too much larger than the split-orbit splitting Δ. Such is the case for InGaAsP at emission wavelengths of $1 \cdot 3 \, \mu m$ ($E_G = 0 \cdot 95 \, eV$, $\Delta \simeq 0 \cdot 25 \, eV$) and $1 \cdot 55 \, \mu m$ ($E_G = 0 \cdot 8 \, eV$, $\Delta \simeq 0 \cdot 3 \, eV$). The relative magnitudes of the CHCC and CHSH processes for InGaAsP have been calculated as approximately the same (within a factor of 2) when parabolic bands are assumed (Dutta and Nelson, 1982; Chiu et al., 1982). However, when more realistic band structures are included, the calculations for $1 \cdot 3 \, \mu m$ material show that the CHSH process dominates by at least an order of magnitude (Sugimura, 1983) although there is also a possibility that the phonon-assisted CHCC process may be the most probable transition (Haug, 1983). Other Auger transitions, such as the CHHH or CHLH processes, are very much less probable for most semiconductors of interest for lasers (Sugimura, 1982). Estimates of the absolute magnitudes of the Auger coefficients A_1, A_2 in eq. 3.5 are difficult in view of uncertainty associated with the quantum-mechanical transition probability (Burt, 1982). Measurements of carrier lifetimes in $1\cdot3\,\mu m$ lasers indicate $A_2 \simeq 10^{-28}\,cm^6\,s^{-1}$ for p-doped material (Su *et al.*, 1982); the total coefficient $(A_1 + A_2)$ for undoped material appears to be less than $10^{-29}\,cm^6\,s^{-1}$ (Su *et al.*, 1982a) although this result has been the subject of controversy (Thompson, 1983; Olshansky *et al.*, 1983, 1984; Yevick and Streifer, 1983). A different measurement on unintentionally doped material yielded a value of $2\cdot3 \pm 1 \times 10^{-29}\,cm^6\,s^{-1}$ (Sermage *et al.*, 1983). For $1\cdot58\,\mu m$ InGaAsP, an Auger coefficient of $4 \times 10^{-29}\,cm^6\,s^{-1}$ has been reported (Asada and Suematsu, 1982).

3.2 Gain and loss

The gain per unit length $g(E)$ is related directly to the rate of stimulated emission defined in the previous Section:

$$g(E) = \frac{\pi^2 c^2 \hbar^3}{E^2 N^2} r_{st}(E) \tag{3.6}$$

where N is the refractive index and the other symbols have their usual meanings. Thus the factors influencing the gain spectrum are the transition probability, the density of states functions, and the occupation probabilities. In particular, it follows from eqn. 3.2 that the gain is only positive for photon energies less than the separation of quasi-Fermi levels, $F_i - F_j$. For a rigid k-selection rule between parabolic bands the low-energy side of the spectrum increases as $(E - E_G)^{1/2}$ and the high-energy side decreases as $(f_i - f_j)$. In the lasing situation, however, the high densities of carriers usually lead to some relaxation of the k-selection rule. Partial k-selection, as characterised by a carrier relaxation time, has been invoked for both GaAs (Yamada and Suematsu, 1978; Zee, 1978) and InGaAsP (Asada *et al.*, 1981); the extreme case of no k-selection has also been applied to these materials (Lasher and Stern, 1964; Osinski and Adams, 1982). For heavily doped semiconductors a sophisticated model of transitions involving band tails with no k-selection was developed first by GaAs by Casey and Stern (1976) and later applied to $1\cdot3\,\mu m$ InGaAsP by Dutta (1980). Fig. 3.4 shows gain spectra for different carrier concentrations, as calculated from the latter model. The photon energy corresponding to peak gain shifts to higher values as the carrier concentration is increased; also the low-energy cut off moves to lower values as a consequence of band-tail effects.

Measurements of gain spectra have been performed by a variety of methods. The most commonly used technique is that due to Hakki and Paoli (1975) in which the gain is deduced from the modulation of the spontaneous emission spectrum associated with Fabry–Perot resonances in a laser structure below

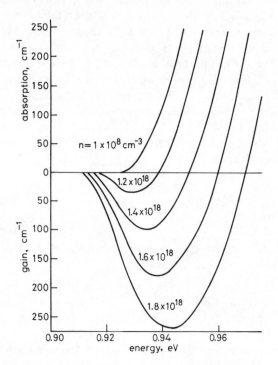

Fig. 3.4 *Calculated gain spectra for 1·3 μm InGaAsP (after Dutta (1980))*
$T = 297$ K
Acceptor concentration = 2×10^{17} cm^{-3}
Donor concentration = 2×10^{17} cm^{-3}

threshold. Results obtained in this way have been reported for InGaAsP at 1·3 μm (Walpole *et al.*, 1981; Dutta and Nelson, 1982*a*), at 1·55 μm (Westbrook, 1986) and at 1·6 μm (Stubkjaer *et al.*, 1981). However, this method is limited to gain values not exceeding threshold, provides only the central part of the net gain spectrum (near its maximum) and contains some uncertainly in the interpretation of the results as it requires knowledge of the internal loss of the laser. An alternative method, developed by Henry *et al.* (1980), utilises the relations (eqns 3.3 and 3.6) to calculate gain spectra from measurements of spontaneous emission emitted perpendicular to the active layer (and thus free from Fabry–Perot resonances and with little subsequent reabsorption). The absolute magnitude of the gain for 1·3 μm InGaAsP was found by measuring absorption spectra on epitaxial layers of identical composition to that used for device fabrication (Henry *et al.*, 1983). A third technique which has been applied to 1·3 μm material involves measurement of the amplification of spontaneous emission generated by optical pumping over a region of variable length (Goebel *et al.*, 1979). A variation of this technique, using electrical excitation on a device with segmented contacts, has also been employed for 1·3 μm quaternary (Prince *et al.*, 1982). A comprehensive review of gain measurements for InGaAsP has been given by Göbel (1982).

A simple empirical formula for the gain is obtained by assuming a linear dependence on electron concentration and a parabolic variation with wavelength λ:

$$g = a(n - n_o) + b(\lambda - \lambda_p)^2 \tag{3.7}$$

where b is related to the gain spectral width, n_o is the 'transparency' electron concentration (for which $g = 0$ at $\lambda = \lambda_p$), and a is the gain coefficient. A further refinement is to include the shift of peak wavelength with n, which is clearly important in Fig. 3.4, in the form:

$$\lambda_p = \lambda_o + (n - n_o)\frac{d\lambda}{dn} \tag{3.8}$$

By fitting eqns. 3.7 and 3.8 to measured gain spectra for $\lambda = 1\cdot55\,\mu\text{m}$ (Westbrook, 1986), values of $a = 2\cdot7 \times 10^{-16}\,\text{cm}^2$, $n_o = 1 \times 10^{18}\,\text{cm}^{-3}$, $b = 0\cdot15\,\text{cm}^{-1}\,\text{nm}^{-2}$ and $d\lambda/dn = -2\cdot7 \times 10^{-17}\,\text{nm cm}^3$ are obtained. Similar empirical formulae have been used for 1·3 μm (Lee *et al.*, 1982) (fitted to the computed results of Dutta (1980)) and for 1·6 μm (Stubkjaer *et al.*, 1981) (fitted to measured gain spectra).

Turning to the question of loss, we consider next free-carrier absorption occurring in a laser as a consequence of the presence of high densities of electrons and holes. For the conduction band the process is clearly an intraband transition involving phonon participation, and is therefore a relatively small effect; the absorption coefficient $\simeq 5 \times 10^{-18}n\,\text{cm}^{-1}$ (with n in cm^{-3}) for GaAs (Kressel *et al.*, 1969). For the valence band, however, there is a stronger absorption arising from transitions between the spin-orbit split-off band and the heavy-hole band, as illustrated in Fig. 3.5. The importance of this loss mechanism for long-wavelength lasers was first recognised by Adams *et al.* (1980) in the context of the temperature dependence of threshold and efficiency (see Section 3.6). Their measurements, and the subsequent ones of Asada and Suematsu (1983), indicate a loss of about $100\,\text{cm}^{-1}$ in 1·6 μm lasers at room temperature. Calculated values for GaInAs are around $70\,\text{cm}^{-1}$ at 1·6 μm and $24\,\text{cm}^{-1}$ at 1·3 μm (Henry *et al.*, 1983a) for $p = 10^{18}\,\text{cm}^{-3}$. For 1·55 μm InGaAsP, by contrast, Sugimura (1981) calculated a loss of about $10\,\text{cm}^{-1}$. Henry *et al.* (1983a) also made direct measurements on GaInAs with $p = 6\cdot2 \times 10^{18}\,\text{cm}^{-3}$; scaling their results to $p = 10^{18}\,\text{cm}^{-3}$, they found losses of $25\,\text{cm}^{-1}$ at 1·6 μm and $13\,\text{cm}^{-1}$ at 1·3 μm, considerably below their theoretical prediction. Mozer *et al.* (1982) observed radiation at about 1·3 eV in 1·3 μm lasers, originating from transitions between the conduction band and the split-off valence band, thus indicating that this valence band is populated by holes either from intervalence band absorption or the CHSH Auger process discussed in Section 3.1. These

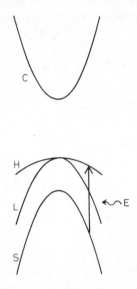

Fig. 3.5 *Inter-valence band absorption of band-gap radiation*

authors (Mozer *et al.*, 1983) also measured the absorption at 1·3 μm caused by free carriers induced by an intense Nd–YAG laser (1·06 μm) in InGaAsP epitaxial layers. Their results for carrier concentrations of order 10^{18} cm^{-3} show a strong increase of absorption coefficient with temperature above 200 K and values of about 400 cm^{-1} at room temperature. Casey and Carter (1984) have measured intevalence band absorption in InP as a function of hole concentration p (in cm^{-3}), and find the absorption coefficient $= 1·4 \times 10^{-17} p$ cm^{-1} at 1·3 μm and $2 \times 10^{-17} p$ cm^{-1} at 1·5 μm.

3.3 Transverse confinement

From eqn. 3.2 we know that to achieve stimulated emission there must be a stronger probability of finding electrons in the upper energy levels than in the lower, a condition usually referred to as 'population inversion'. In order to bring about this condition, early injection lasers incorporated a strongly forward-biased p–n junction; the quasi-Fermi level separation $(F_i - F_j)$ is then equal to the voltage dropped across the junction multiplied by the electron charge. However, for present-day lasers a more sophisticated structure is employed — the 'double heterostructure' illustrated in Fig. 3.6. This structure provides a well defined region − the central 'active' layer − where the recombination occurs, and carrier wastage by diffusion to other regions is prevented by potential barriers at the hetero-interfaces. The most common devices of this kind consist of a GaAs active layer between *n*- and *p*-type layers of AlGaAs, so that the outer layers are of larger band gap and therefore transparent to the emitted radiation. A comprehensive treatment of heterostructure lasers is given in the book by Casey and Panish (1978).

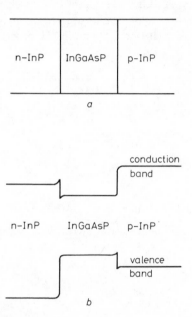

Fig. 3.6 *The InGaAsP/InP double heterostructure*
a layer structure
b band diagram under forward bias

The band gap of GaAs at room temperature is about 1·43 eV, so that the emission wavelength lies at about 0·9 μm. Fibre losses and material dispersion are minimised in the range 1–1·6 μm with especial emphasis on the wavelengths 1·3 and 1·55 μm, and to make sources for this range means using materials other than the AlGaAs/GaAs system. The most promising materials system is the quaternary InGaAsP using InP cladding layers to produce the double hetero-structure, and it is this system which is shown in Fig. 3.6. Longer wavelengths imply smaller band gaps, and this places one restraint on the choice of material. A second restraint comes from the necessity of making strain-free hetero-junctions with good-quality substrates. The requirement of minimising strain effects arises from a desire to avoid interface states and to encourage long-term device reliability, and this imposes a lattice-matching condition on the materials used. The two constraints of band gap and lattice match mean that quaternary semiconductors of the form $A_x B_{1-x} C_y D_{1-y}$ are needed so that the two parameters x and y may be chosen to satisfy these conditions. Fig. 3.7 summarises the possible materials systems for long-wavelength operation on a plot of lattice constant versus emission wavelength/energy gap. The InGaAsP system spans the wavelength range from $0.92\,\mu m$ (InP) to $1.67\,\mu m$ (InGaAs). Alternative materials systems include GaAlInAs lattice-matched to InP substrates ($0.87\,\mu m$ to $1.65\,\mu m$) or GaAlAsSb on GaSb substrates. Lasers have been fabricated in the latter system with operating wavelengths in the range 1.5–$1.8\,\mu m$ (Dolginov *et al.*, 1981), although threshold currents are very high at the $1.5\,\mu m$ end of this range, probably due to a strong Auger CHSH process (Burt, 1981) and/or inter-valence band absorption.

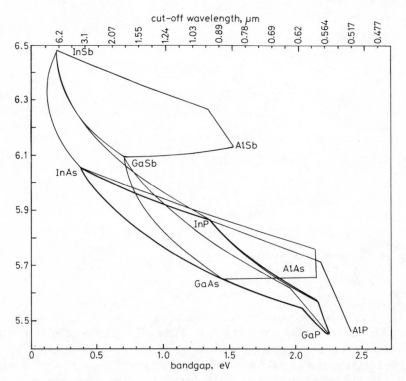

Fig. 3.7 *Lattice constant versus energy-gap/emission-wavelength for some III-V alloy semiconductors*
Data from Glisson *et al.*: *J. Electron. Mat.*, 1978, **7** (1)

The epitaxial layers of the heterostructure laser help to improve its performance in two ways: by confining the carriers to the active layers and simultaneously confining the electromagnetic radiation to this region. This latter effect stems from the fact that the narrower-gap active material has a higher refractive index than that of the wider-gap passive layers. For example, for

GaAs at a wavelength of $0.85\,\mu m$, the refractive index is 3.6, whilst for $Al_{0.3}Ga_{0.7}As$ the value is about 3.4. Thus the double heterostructure forms a symmetric dielectric slab waveguide. For InGaAsP lattice matched to InP there is still some uncertainty over refractive indices (see, for example, Burkhard (1984) and references therein), but values of 3.54 at $1.3\,\mu m$ and 3.57 at $1.55\,\mu m$ are reasonably well established. For cladding layers of InP, the corresponding indices are (Pettit and Turner, 1965) 3.20 at $1.3\,\mu m$ and 3.17 at $1.55\,\mu m$.

From the viewpoint of understanding laser behaviour the significant waveguide parameter is the confinement factor Γ, defined as the ratio of power confined in the core to the total cross-sectional power. Values of this parameter can be calculated as a function of a normalised active-layer thickness, $D = (2\pi d/\lambda)(N_1^2 - N_2^2)^{1/2}$, where d is the true thickness, λ the wavelength, and N_1, N_2 the refractive indices of core and cladding waveguide layers, respectively. Fig. 3.8 includes a plot of Γ versus D (broken line) for the lowest-order transverse mode of a double heterostructure. At large values of D the confinement factor approaches unity since the wave is extremely well-confined; as the value of D decreases, the wave spreads into the cladding regions and Γ decreases, tending to zero as $D \to 0$. A useful approximation for the confinement factor, accurate to within 1.5%, is given by Botez (1978):

$$\Gamma \simeq \frac{D^2}{2 + D^2} \tag{3.9}$$

Clearly, for the wave to experience as large a gain as possible, a high value of Γ is desired, and this can be achieved by increasing D. However, this also leads to the possibility that higher-order transverse modes will be permitted; to ensure single transverse-mode behaviour, D must be less than π. For the refractive indices given above this corresponds to active layer thicknesses of $0.43\,\mu m$ and $0.47\,\mu m$ at wavelengths of $1.3\,\mu m$ and $1.55\,\mu m$, respectively.

The far-field emission pattern for the fundamental mode of the symmetric slab waveguide has a beamwidth $\theta_\perp$ which is given to a good approximation by the formula (Botez and Herskowitz, 1980):

$$\theta_\perp = \frac{0.65D\,(N_1^2 - N_2^2)^{1/2}}{1 + 0.15(1 + N_1 - N_2)D^2} \tag{3.10}$$

This result is claimed to be accurate within 3% for $D \leqslant 2$, which corresponds to $d \leqslant 0.3\,\mu m$ for GaInAsP at $1.55\,\mu m$. Narrow beamwidths can thus be obtained from lasers with very thin active layers. For example, a value of $23°$ has been obtained on a $1.3\,\mu m$ laser with $d = 0.05\,\mu m$ (Itaya *et al.*, 1979); for these parameters eqn. 3.10 predicts $\theta_\perp = 20°$ using the refractive indices given earlier.

A number of structures have been developed with the object of separating the functions of optical and carrier confinement, which are synonymous in the double heterostructure; Fig. 3.9 shows the refractive-index profiles of three of these. Fig. 3.9*a* is the large optical cavity (LOC) four-layer structure proposed originally for high-power lasers (Lockwood *et al.*, 1970).

The material composition of the layers is chosen such that recombination of electrons and holes occurs only in the highest-index layer of thickness *d*, whilst

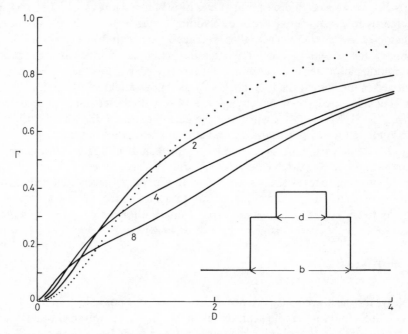

Fig. 3.8 *Confinement factor Γ versus normalised active layer thickness D for five-layer symmetric waveguides*
The ratio of index differences (core/inner cladding: inner/outer cladding) is 1 : 3, and the labelling parameter gives the ratio of inner cladding + core thickness (*b*) to core thickness (*d*) (see inset). The dotted line corresponds to the equivalent three-layer slab waveguide

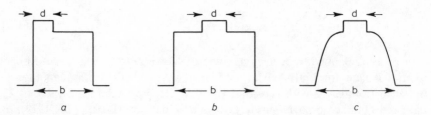

Fig. 3.9 *Refractive-index profiles for three laser structures*
a Large optical cavity (LOC)
b Localised gain region (LGR) or separate confinement heterostructure (SCH)
c Graded-index waveguide with separate confinement heterostructure (GRIN-SCH)

the optical field can spread throughout the central two layers of total thickness *b*. Although early versions of this structure permitted several transverse modes to propagate, single-mode operation has been demonstrated by suitable choice of dimensions and refractive-index steps (Paoli *et al.*, 1973). Fig. 3.9*b* shows the refractive-index profile of a five-layer structure known variously as the localised-gain-region (LGR) laser (Thompson and Kirkby, 1973) or the separate-confinement-heterostructure (SCH) laser (Panish *et al.*, 1973). The active recombination region of this structure is the central high-index layer, whilst the optical field is allowed to spread throughout the middle three layers. Fig. 3.8 shows three curves of confinement factor Γ versus normalised width *D* for the waveguide of Fig. 3.9*b*; the ratio of index differences has been taken as 1 : 3, and the labelling parameter on the curves gives the ratio of the optical guide thickness *b* to the active layer thickness *d*. For small values of *D* the confinement for the LGR/SCH structure is considerably better than for the DH device (dotted line).

Fig. 3.9*c* shows a variant of the conventional LGR/SCH laser, namely the graded-index-waveguide separate-confinement heterostructure (GRIN–SCH) device (Tsang, 1981). Compared with the five-layer structure of Fig. 3.9*b*, the GRIN–SCH laser offers three additional features: (i) the ability to control the far-field distribution of radiation from the waveguide, (ii) an enhanced suppression of higher-order modes, and (iii) a shift of active layer thickness for cut-off of the first-order mode to higher values. As a consequence, GRIN–SCH lasers have very low threshold current densities ($\sim 500 \, \text{A/cm}^2$ for the AlGaAs system) and can give narrow output beams of Gaussian far-field distribution when the refractive-index profile is parabolic.

3.4 Quantum wells

When the active-layer thickness of a double heterostructure is reduced to the order of the carrier de Broglie wavelength, then new effects, dissimilar from the characteristics of the bulk material, become significant. These quantum-size effects (QSE) arise from the confinement of carriers to the finite potential wells formed by conduction and valence band edges. Fig. 3.10 illustrates the situation for a 100 Å layer of $Ga_{0.47}In_{0.53}As$ between barriers of $Al_{0.48}In_{0.52}As$, these compositions being lattice-matched to InP. For this case, the conduction-band discontinuity at the heterojunction is $0.5 \pm 0.05 \, \text{eV}$ (People *et al.*, 1983) and the corresponding valence-band step is $0.2 \pm 0.05 \, \text{eV}$. The motion of carriers in the GaInAs layer is quantised for the components perpendicular to the well, giving rise to the characteristic energy levels shown in Fig. 3.10. Three levels are predicted for electrons and five for heavy holes, the difference arising from the large discrepancy between the effective masses of these carriers.

In directions parallel to the hetero-interfaces, on the other hand, the carriers are free to move. This means that the density of states for the carriers will be that of a two-dimensional system rather than the familiar case of three-dimensional motion. The difference is illustrated in Fig. 3.11, for the case of the electrons in the 100 Å well shown in Fig. 3.10. The 3-D density of states is given by the curved line with the usual square-root dependence on energy, whilst the 2-D case is given by the 'staircase' with each step occurring as a new sub-band is reached (corresponding to the discrete levels shown in Fig. 3.10). A comprehensive discussion of quantum-well properties arising from this density-of-states behaviour is given by Dingle (1975); a review of quantum-well lasers has been presented by Holonyak *et al.* (1980).

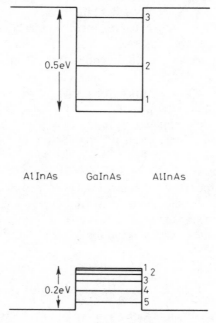

Fig. 3.10 *Band diagram for a GaInAs–AlInAs quantum well, showing the energy eigenvalues of confined particle states*

It is clear from Fig. 3.10 that the lowest-energy transition between electron and heavy-hole states is at a photon energy significantly higher than the band gap. For the 100 Å well illustrated the photon energy will in fact be about 0·81 eV (1·53 μm) as compared to 0·75 eV for the energy-gap. Of course, this photon energy will depend on the well width and will increase to higher values as the width decreases. The photon energy and the corresponding levels for confined carriers depend also on the height of the potential barriers, and hence on the materials used for the well and the barriers.

The theory for the stimulated emission rate, and hence gain, for quantum wells follows along the same lines as discussed for bulk material in Sections 3.1 and 3.2. However, the 'staircase' density of states in the quantum well means that the gain spectrum will be somewhat narrower than for the bulk case, since the low-energy side of the spectrum is determined by the form of the density of states. For the quantum well there is a selection rule which permits only transitions between equivalent sub-bands in the conduction and valence bands.

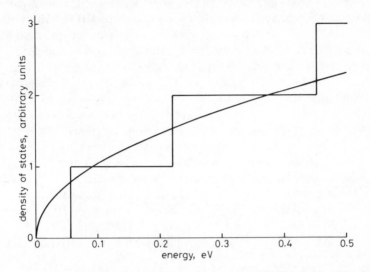

Fig. 3.11 *Density of states versus energy for conduction-band electrons in a GaInAs–AlInAs quantum well ('staircase')*
The curved line gives the corresponding density of states distribution for bulk GaInAs

Hence gain calculations which assume a rigid k-selection rule result in spectra with an abrupt low-energy cut-off at the energy separation between the lowest confined particle states (Kasemset *et al.*, 1983). Gain spectra measured by the method of Hakki and Paoli (1975) on GaAs–AlGaAs quantum wells, however, do not exhibit a sharp low-energy cut-off (Dutta *et al.*, 1983; Kobayashi *et al.*, 1983). This discrepancy may be explained either by the existence of broadening mechanisms which 'soften' the low-energy side of the spectrum, or by a break-down of the k-selection rule, (Hess *et al.*, 1980; Landsberg *et al.*, 1984). Spectral characteristics of photopumped quantum-well lasers (Holonyak *et al.*, 1980) indicate that longitudinal optical phonons participate in the recombination process. However, Sugimura (1983*a*) has theoretically evaluated the phonon-assisted gain for AlGaAs quantum-well lasers and has shown that at room temperature it is smaller than the direct transition gain by more than an order of magnitude. Phonon-assisted laser operation is therefore expected only under conditions where the cavity loss for direct emission is substantially higher than that for phonon-assisted emission.

As regards Auger recombination in quantum wells, there have been a number of calculations of transition rates, but only one experimental measurement has been published at the time of writing. Chiu and Yariv (1982) have made numerical calculations of the CHCC Auger transition rate for $1\cdot3\,\mu m$ InGaAsP–InP quantum wells. For a $100\,\text{Å}$ well their results show that the CHCC lifetime is about two orders of magnitude larger than for bulk InGaAsP under identical injection conditions; the Auger lifetime decreases rapidly with increasing well width. Sugimura (1983b) has calculated CHSH Auger lifetimes for $1\cdot07\,\mu m$ InGaAsP–InP quantum wells showing some indication of lifetime decrease with increasing well width. However, there is also an undulation of the lifetime for widths around $100\,\text{Å}$, which may be associated with changes of quantisation effects for the spin-orbit split-off valence band. In general the CHSH lifetimes are comparable to the corresponding radiative lifetimes, although the former show a stronger dependence on carrier concentration than the latter. Dutta (1983) has reported calculations of Auger lifetimes for the CHCC, CHSH and CHLH processes in InGaAsP–InP quantum wells of width $200\,\text{Å}$. The results are claimed only to be order-of-magnitude estimates in view of the uncertainties of band structure; they show the CHCC and CHSH rates to be of the same order whilst the CHLH rate is about two orders of magnitude less. At room temperature, CHCC lifetimes in the range 6–$100\,\text{ns}$ for $1\cdot3\,\mu m$ quantum wells are predicted for $n = 10^{18}\,\text{cm}^{-3}$, the values decreasing rapidly with increasing temperature; at $1\cdot5\,\mu m$ the corresponding range of CHCC lifetimes is 1–$50\,\text{ns}$.

An overall assessment of the wide range of numerical results on Auger recombination cited above is difficult since many authors do not include sufficient mathematical detail to make comparisons possible. By contrast, Smith *et al.* (1983) give systematic derivations of algebraic expressions for the CCCH process in a quantum well with the carriers involved remaining in the lowest electron and hole sub-bands. For this model with equal concentrations of holes and electrons, and provided the wells are not too narrow, the ratio of Auger recombination rate in conventional (bulk) lasers to that in quantum-well devices is of the order of $(E_a/kT)^{1/2}$, where E_a is the activation energy of the bulk Auger process (Smith *et al.*, 1984; see also Tsang, 1984). Since E_a is typically of order $100\,\text{meV}$ in quaternary material, it follows that at room temperature the quantum-well Auger rate is predicted to be only a factor of two smaller than the corresponding rate in bulk material. This prediction is supported by an experimental measurement of Auger rates in GaInAs–AlInAs multiple quantum wells and in bulk GaInAs (Sermage *et al.*, 1986) which showed no significant difference in the two rates.

Dutta (1983) has also estimated the magnitude of intervalence band absorption in InGaAsP–InP quantum wells of width $200\,\text{Å}$. At room temperature with $n = 10^{18}\,\text{cm}^{-3}$, he finds an absorption coefficient of $2\,\text{cm}^{-1}$ at $1\cdot55\,\mu m$ for transitions between the lowest-heavy-hole and split-off sub-bands. When other sub-bands are populated, the intervalence band absorption may be larger by as much as a factor of 5.

The narrow active-layer widths of quantum wells imply that the confinement of radiation within a single well is very poor and the confinement factor Γ is extremely small, as a glance at Fig. 3.8 will confirm. One approach to improving the optical confinement is to use multiple quantum wells (MQWs) separated by thin barrier layers. For GaAs–AlAs wells Suzuki and Okamoto (1983) have shown that when the barriers are less than 40–50 Å wide the refractive index of an MQW structure is determined by the average AlAs mole fraction, as suggested by Streifer *et al.* (1979). For larger barrier widths the contribution of the quantised well states to the refractive index results in somewhat lower index values (Leburton and Hess, 1983). However, in both cases there is a substantial increase in the confinement factor for MQW as compared to single quantum-well structures, and this has been utilised in the development of MQW AlGaAs lasers with very low ($\sim 250\,\text{A/cm}^2$) threshold current densities (Tsang, 1981*a*). A second solution to the problem of poor optical confinement (and electron capture) in single quantum wells is provided by use of SCH and GRIN–SCH structures; this has also led to the achievement of extremely low-threshold AlGaAs lasers in a number of laboratories (Tsang, 1982; Kasemset *et al.*, 1982; Hersee *et al.*, 1982; Burnham *et al.*, 1982).

Long-wavelength quantum-well lasers have been demonstrated recently. Yanase *et al.* (1983) have made 1·3 μm GaInAsP–InP MQW lasers by hydride-transport VPE with threshold current densities of about 1000 A/cm². Temkin *et al.* (1983) have reported 1·5–1·6 μm GaInAs–AlInAs MQW lasers grown by MBE with a lowest threshold of 2400 A/cm². Using MOCVD growth, Razeghi and Duchemin (1983) have fabricated 1·5 μm GaInAs–InP MQW lasers with thresholds down to 1700 A/cm². With the same materials system, but using MBE growth, Tsang (1984) has reported 1·53 μm MQW lasers with a best threshold of 2700 A/cm².

3.5 Laser cavities

The simplest semiconductor laser structure employs a Fabry–Perot (FP) cavity formed by cleaved facets at each end of the device (Fig. 3.12*a*). The high refractive index of the semiconductor gives sufficient reflectivity (typically 30%) at the facets to produce a resonant cavity. Denoting the length of the cavity by L and the facet reflectivities* by R_1, R_2, the transmission coefficient T for the cavity is given by:

$$T = \frac{(1 - R_1)(1 - R_2)e^{GL}}{(1 - \sqrt{R_1 R_2}\,e^{GL})^2 + 4\sqrt{R_1 R_2}\,e^{GL}\sin^2\left(\dfrac{2\pi NL}{\lambda}\right)} \quad (3.11)$$

---

* Note that here, and in what follows, we assume plane-wave reflectivities R_1, R_2. In fact, these should be modal reflectivities, but this refinement is neglected here since it represents only a second-order correction to the results.

where N is an equivalent refractive index for the guided mode, defined as the ratio of propagation constant in the waveguide to that in free space, and G is the modal gain per unit length. Denoting the material gain by g (as in Section 3.2), the loss in the active layer by α_{act}, and the loss in the passive cladding layers by α_{cl} (see Fig. 3.12a), it follows that the modal gain G is given by

$$G = \Gamma(g - \alpha_{act}) - (1 - \Gamma)\alpha_{cl} \tag{3.12}$$

where Γ is the optical confinement factor defined in Section 3.3.

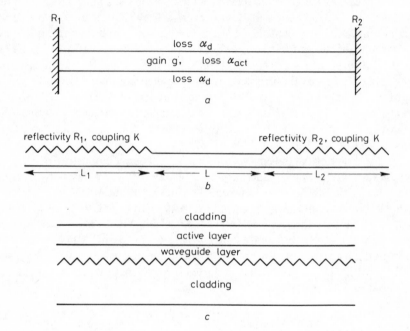

Fig. 3.12 *Laser cavities*
    a Fabry–Perot (DH structure)
    b Distributed Bragg reflector
    c Distributed feedback (LOC structure)

Fig. 3.13 shows a plot of the transmission coefficient T from eqn. 3.11 versus wavelength λ for a laser just below threshold. The cavity resonances occur for integer values of the quantity $2LN/\lambda$, and thus the spacing between resonances is given by

$$\delta\lambda = \frac{\lambda^2}{2LN_g} \tag{3.13}$$

where N_g is the group effective index ($N - \lambda dN/d\lambda$). For a cavity length L of $300\,\mu m$, eqn. 3.13 yields values for the mode spacing of 0·3, 0·7 and 1·0 nm at operating wavelengths of 0·85, 1·3 and 1·55 μm, respectively.

The distributed-Bragg-reflector (DBR) laser, illustrated schematically in Fig. 3.12*b*, replaces the mirrors of the FP cavity by gratings. The pitch Λ of the grating is related to the wavelength by $\Lambda = p\lambda/2N$, where p is an integer representing the order of the grating. For example, at a wavelength of $1.55\,\mu$m,

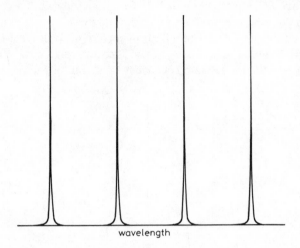

wavelength

Fig. 3.13 *Fabry–Perot cavity transmission versus wavelength for a laser just below threshold*

a first-order grating has a pitch of $0.23\,\mu$m and a second-order grating has a pitch of $0.46\,\mu$m. Higher orders than the second lead to strong coupling to radiation modes and are therefore only rarely used, and then with specific structures (grating-coupled lasers) to produce output from the surface of the laser. The reflection R_1 for a grating of length L_1 is given by

$$R_1 = \frac{|\kappa|^2}{|\alpha + i\Delta\beta + \gamma \coth(\gamma L_1)|^2} \tag{3.14}$$

where $\gamma^2 = \kappa^2 + (\alpha + i\Delta\beta)^2$, α is the amplitude loss coefficient, κ is the grating coupling constant, and $\Delta\beta$ is the deviation of propagation constant from the Bragg condition

$$\Delta\beta = \frac{2\pi N}{\lambda} - \frac{p\pi}{\Lambda} \tag{3.15}$$

The reflection coefficient R_1 from eqn. 3.14 is plotted in Fig. 3.14 as a function of $\Delta\beta L_1$, for the case $\alpha = 0$ and $\kappa L_1 = 2$. The wavelength dependence of the reflectivity means that the DBR laser has improved discrimination against unwanted longitudinal modes over the FP laser; we shall return to this later in our discussion of spectra.

The distributed-feedback (DFB) laser shown in Fig. 3.12*c* is somewhat similar to the DBR, but the grating is now incorporated along the length of the gain region rather than outside it. To avoid non-radiative recombination associated with defects introduced by the grating fabrication process, it is necessary to separate the corrugations from the active layer whilst still allowing them to interact strongly with the optical field. Usually this is achieved by the use of an SCH/LOC structure, as discussed in Section 3.3, where the grating is separated from the active layer by a layer of lower refractive index. Thus the grooves interact with the evanescent tail of the optical distribution to provide feedback along the length of the laser. The reflection coefficient for a DFB laser is given by an expression similar to eqn. 3.14, but with α now replaced by a negative term corresponding to the modal gain coefficient ($-G/2$ in our notation).

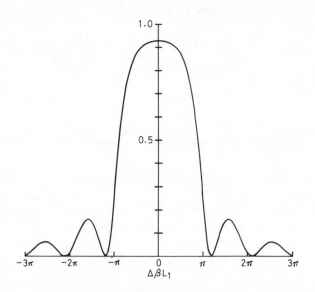

Fig. 3.14 *Grating-reflectivity spectrum from eqn. 3.14 for $\alpha = 0$ and $KL_1 = 2$*

3.6 Threshold and efficiency

The threshold relation for an FP laser is given by the condition that the denominator of the RHS of eqn. 3.11 vanishes at the resonant wavelengths. Combining this result with eqn. 3.12 for G yields

$$g = \alpha_{act} + \left(\frac{1-\Gamma}{\Gamma}\right)\alpha_{cl} + \frac{1}{2\Gamma L}\ln\left(\frac{1}{R_1 R_2}\right) \qquad (3.16)$$

This equation states that the material gain supplied to that part of the mode occupying the active layer is exactly balanced by the loss suffered by the proportions of the mode in the active and cladding layers, together with the end loss through the cavity mirrors. This result is only approximately true, since it neglects the small amount of spontaneous emission coupled into the mode; in reality, the threshold value of g will be a little less than the RHS of eqn. 3.16, the difference being given by the spontaneous emission.

For a DBR laser the threshold condition is again given by an equation of the same form as eqn. 3.16, but with two significant differences: (i) the reflectivities R_1, R_2 are now functions of wavelength, see e.g. eqn. 3.14, and (ii) the end-loss term must be modified by the inclusion of a coupling efficiency C for coupling between the active region and each grating. The efficiency C appears as a multiplier of $R_1 R_2$ in the final term of eqn. 3.16 for the DBR case (Stoll, 1979; Suematsu *et al.*, 1983).

In analogy with the FP case, the threshold relation for the DFB laser is given by the condition that the denominator of the transmission coefficient be zero (Kogelnik and Shank, 1972). Since this denominator is, in fact, identical to that of the corresponding reflection coefficient, then threshold corresponds to the zero of the denominator of eqn. 3.14 with the appropriate interpretation of α $(-G/2)$ as noted earlier. In general, the complicated form of the denominator in eqn. 3.14 means that threshold gains and resonant wavelengths must be computed numerically for the DFB laser. However, approximate analytic results are available for the limiting cases of high and low gain ($\kappa \ll G$ and $\kappa \gg G$, respectively) (Kogelnik and Shank, 1972).

The nominal current density j_{th} required at lasing threshold is proportional to the total spontaneous emission rate R_{sp}, as given in eqn. 3.4:

$$\frac{j_{th}}{ed} = \frac{Bnp}{\eta_i} \tag{3.17}$$

where η_i is the internal quantum efficiency, defined as the ratio of radiative recombination rate to the total (i.e. radiative plus non-radiative) rate. The RHS of eqn. 3.17 can be expressed in terms of the gain at threshold by using the gain-carrier concentration relation (eqn. 3.7) evaluated at peak gain ($\lambda = \lambda_p$), and using the charge-neutrality condition to relate p and n. The result can usually be expressed as a linear current-gain relation over the range of interest, in the form

$$j_{th} = \frac{d}{\eta_i \beta_o} \left[\alpha_o + \alpha_{act} + \left(\frac{1 - \Gamma}{\Gamma} \right) \alpha_{cl} + \frac{1}{2\Gamma L} \ln \left(\frac{1}{R_1 R_2} \right) \right] \tag{3.18}$$

where an FP cavity has been assumed and the threshold condition (eqn. 3.16) has been used. The parameter α_o in eqn. 3.18 is the loss coefficient at transparency, and β_o is a material-dependent scale factor. Numerical calculations by Dutta (1980, 1981) for $1\cdot3\,\mu$m InGaAsP have confirmed the linear approximation of eqn. 3.18 and predicted values for α_o and β_o for various doping levels and temperatures. Botez (1981) has surveyed the experimental evidence for eqn. 3.18 in quaternary lasers and has chosen mean values of $\eta_i \beta_o = 28\cdot5\,\mathrm{cm}\,\mu\mathrm{m/kA}$ and $\alpha_o = 50\,\mathrm{cm}^{-1}$, and assumed they do not vary significantly with wavelength for the quaternary at room temperature.

The dependence of threshold current density on the waveguide aspect of the device enters eqn. 3.18 via the thickness d and the confinement factor Γ. Far from cut-off of the fundamental transverse mode, the field is well-confined to the central active layer and the threshold simply increases linearly with the width of this layer. Nearer cut-off the field spreads into the passive regions, and this feature dominates the behaviour until at cut-off the field spreads uniformly throughout all space, approaches zero, and the threshold becomes very large. Thus there is always an optimum value of thickness d, between these limits, to give minimum threshold current density. For the case where the end loss dominates over the internal losses the threshold dependence is largely determined by the ratio d/Γ in eqn. 3.18, and this has an optimum when the normalised thickness D is 1·42 (Unger, 1971). In practice the optimum layer thickness for quaternary lasers is around 0·15 μm (Nahory and Pollack, 1978; Botez, 1981) when the value of threshold current density is approximately 1 kA/cm^2.

It has become customary to represent the temperature dependence of threshold current by an exponential function, i.e. $j_{th} \propto \exp (T/T_o)$, where T_o for InGaAsP lasers has been found experimentally to be in the range 50–80 K for $T > 250$ K. This temperature sensitivity is significantly higher than is the case for GaAs–AlGaAs lasers, where T at room temperature is usually in the range 120–180 K. Looking at eqn. 3.18, we may identify the origins of the temperature dependence of j_{th}, as follows:

(i) Thermal broadening of carrier distributions via the temperature dependence of the Fermi–Dirac function; this fundamental mechanism introduces a temperature dependence into the quantities α_o and β_o of eqn. 3.18. Theoretical calculations for GaAs based on this effect are able to account for the experimental results (Casey and Panish, 1978), but this is not the case for quaternary lasers (Dutta, 1981).

(ii) Non-radiative recombination via temperature-activated defect (Horikoshi and Furukawa, 1979); this temperature sensitivity would enter eqn. 3.18 through the internal efficiency η_i. However, no identification of such a defect in InGaAsP has yet been given.

(iii) Auger recombination, also giving rise to a temperature dependence of η_i. As discussed in Section 3.1 there is some uncertainty, both theoretical and experimental, over the magnitudes of the various Auger effects in InGaAsP.

(iv) Inter-valence band absorption, giving rise to a temperature dependence of the term α_{act} in eqn. 3.18. Once again, however, the size of this effect is uncertain, as the discussion in Section 3.2 has demonstrated.

(v) Carrier leakage over a hetero-barrier; whilst not included explicitly in eqn. 3.18, this effect would manifest itself if the nominal current density j_{th} were to be related to the measured current density at threshold. Electron leakage has been measured and found to be a substantial fraction of the total injection current (Chen et al., 1983).

At the time of writing the relative importance of these mechanisms for the temperature sensitivity of quaternary lasers has yet to be clarified, although a recent survey by Casey (1984) has identified (iii) and (v) as the dominant effects. In spite of intense study of this problem, it has proved impossible to obtain high T_o values in these devices other than by the use of structures which include less temperature-dependent current shunts.

The light-output/current characteristic for an injection laser is shown schematically in Fig. 3.15. If we define P as the optical power emitted by the

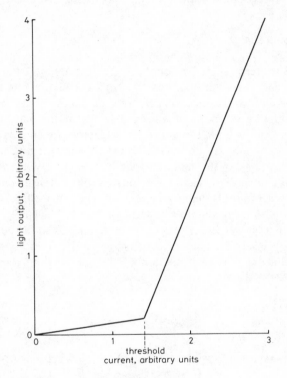

Fig. 3.15 *Schematic of light-output/current characteristic*

device above threshold, then for a well behaved laser the relation to current I and threshold current I_{th} is given by

$$\frac{P}{E} = P_{sp}\eta_i \frac{I_{th}}{e} + \eta_D \frac{(I - I_{th})}{e} \tag{3.19}$$

where P_{sp} is a geometrical factor giving the probability of a spontaneous photon being emitted from a facet. The quantity η_D is the differential quantum efficiency, defined as the ratio of end loss to total gain experienced in the device. From eqn. 3.16, it follows that

$$\eta_D = \frac{\dfrac{1}{2L} \ln\left(\dfrac{1}{R_1 R_2}\right)}{\Gamma\alpha_{act} + (1 - \Gamma)\alpha_{cl} + \dfrac{1}{2L} \ln\left(\dfrac{1}{R_1 R_2}\right)} \qquad (3.20)$$

A more rigorous derivation of this expression for η_D, together with a discussion of earlier literature on the subject, is given by Nash and Hartman (1979). The principal temperature dependence of the differential efficiency arises from inter-valence band absorption via the term α_{act} in eqn. 3.20 (Adams *et al.*, 1980).

3.7 Lateral confinement

In Section 3.3 the term 'transverse confinement' was used to describe the effects of the double heterostructure (and more complex structures) on the motion of carriers and radiation in a direction perpendicular to the hetero-barriers. In the present Section we are concerned with confinement in the plane of the hetero-structure layers, and we reserve the adjective 'lateral' for this direction. In the early GaAs lasers, there was no attempt to produce any form of lateral confine-ment for the carriers or the radiation; the devices were what would now be termed 'broad contact' lasers and suffered from the twin disadvantages of high threshold currents and filamentary behaviour. The latter effect consists of the formation of one or more longitudinal lasing filaments which may switch on and off and exchange power in an uncontrolled manner. The introduction of the stripe-geometry laser (Dyment and D'Asaro, 1967), where a narrow stripe contact is defined by oxide insulation of the surrounding regions, avoided

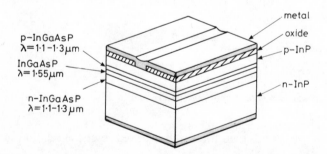

Fig. 3.16 *Oxide-stripe long-wavelength laser structure*

filamentary action and reduced threshold currents considerably. Fig. 3.16 shows the structure of an oxide-stripe long-wavelength laser. With typical device lengths of order $300–500 \, \mu m$ the threshold current of such a device is of the order of 100 mA; radiation is emitted with a far-field width of about 6° parallel to the junction and 50° in the perpendicular direction. Such devices will usually launch power of the order of 1 mW into a multimode optical fibre either by a simple butt coupling or by use of a cylindrical lens to restrict the beam divergence perpendicular to the junction plane.

If we consider the oxide-insulated structure of Fig. 3.16, it is clear that, in the region under the stripe and for some distance on either side due to current spreading and carrier diffusion, there is a region of gain; elsewhere along the junction plane the remaining area will be lossy. The refractive index of these regions with gain and loss will be, to a first approximation, the same. However, the combined effects of free-carrier concentration, photoelasticity associated with the stripe, and temperature variations along the junction will produce either a real dielectric waveguide (for wide stripes $> 15\,\mu$m) or an antiguiding effect (especially in narrow-stripe lasers). In the case of antiguiding, the confinement of the wave occurs by a 'gain-guiding' effect, usually with increased internal losses.

Analyses of these effects can be made by applying the 'effective index' method (see e.g. Buus, 1982) to the device, since the guidance perpendicular to the junction plane is much stronger (relative to the junction plane is much stronger (relative index differences as high as 10%) than that parallel to the junction (relative index difference of order 10^{-3}, gain differences of order 10–$100\,\mathrm{cm}^{-1}$). Hence the three-layer slab problem perpendicular to the junction is first solved at each position along the plane, and the result is then used to define an effective permittivity for an equivalent graded-index complex guide parallel to the junction plane. The result of the analysis may be expressed in terms of observable parameters — spot size and phase front radius of curvature — which may be compared with experiment and thus used to find the relative amounts of gain guiding, real guiding, or antiguiding present in a given device. Such studies have led to a detailed understanding of device behaviour for the stripe-geometry heterostructure laser with implications for mode stability, output linearity and transient behaviour.

In many oxide-insulated devices there are departures from the simple linear dependence of light output on current above threshold as shown in Fig. 3.15 and eqn. 3.19. It is now generally accepted that the phenomenon of 'kinks' in the light/current curve are caused by a combination of gain- and index-guiding effects on the lateral mode profile. These effects are avoided by the use of structures with build-in index guides in the junction plane, such as the buried heterostructure or channelled substrate devices. These structures will be discussed in detail in Chapter 4. Lasers with a deliberate lateral waveguide usually are characterised by narrow near-field distributions, stable optical output, and linear light/current curves up to 10 mW or more. These features are results of the strong dielectric guiding effect and closely controlled dimensions which suppress unwanted higher-order modes and increase the fundamental mode output power. Such devices also show an almost flat frequency response out to beyond 1 GHz with no resonance effects such as are associated with oxide-stripe (or proton-bombarded) structures (see Chapter 7).

3.8 Laser spectra

The gain of semiconductor laser materials is relatively broad when compared with the longitudinal mode spacing or 'selectivity' of the Fabry–Perot cavity, and in practice this generally results in multi-longitudinal mode operation. Compared with typical mode spacings of a few nm, the gain extends over many longitudinal modes (cf. Fig. 3.4). In order to determine the output spectrum a rate equation for each longitudinal mode is used (see eqn. 7.10), and in equilibrium the photon density of each mode i can be shown to be given by

$$S_i = \frac{\beta_i R_{sp}}{\dfrac{\Gamma c}{N} g_i - \dfrac{1}{\tau_p}} \qquad (3.21)$$

where g_i is the mode gain of the ith longitudinal mode and β_i the spontaneous coupling factor. The total output is given by a sum over all modes. In eqn. 3.21, τ_p is the photon lifetime, which is related to loss in the cavity (see eqn. 7.3).

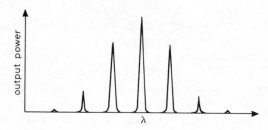

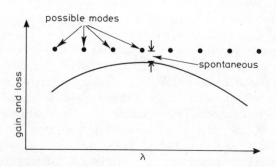

Fig. 3.17 *Schematic of allowed longitudinal modes and gain of a Fabry–Perot laser*
The multimode spectrum is due to the broad gain curve in relation to the longitudinal mode spacing
Loss = gain + spontaneous

Fig. 3.17 shows a schematic diagram of the gain an
spectra for a typical semiconductor laser. From eqn. 3.2
in any mode is inversely proportional to the difference bet\
loss, and this gives rise to the mode spectrum, which is st

Fig. 3.18 shows the output spectrum for two devices of
the cavity length is reduced from 200 μm to 100 μm the
spacing increases (cf. eqn. 3.13) and so for shorter devic   .... fewer
allowed modes in regions of high gain. Thus shorter devices tend to have fewer
modes in their output spectrum. To offset this effect it has been suggested that
the gain is broadened when the carrier concentration is high, as is required in
order to achieve threshold in short lasers.

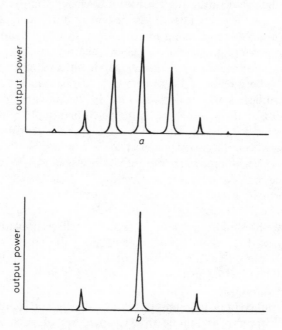

Fig. 3.18 *Schematic diagram of the spectra obtained on devices of different length*
As the length decreases the mode spacing increases, and so at the centre of the gain
spectrum there is a lower density of allowed longitudinal modes
a standard length (typically 200 μm)
b effect of halving length

The main factor governing the operating wavelength of a semiconductor laser
is the position of the peak of the material gain at threshold. Since the gain peak
moves to shorter wavelengths as the carrier density increases (see Fig. 3.4) the
operating wavelength can be varied by altering the gain required at threshold.
This can be achieved by either varying the device length or the reflectivity of the
facets. For example, a range of ~20 nm can be covered by altering the device

from $100 \mu m$ to $300 \mu m$ ($\lambda = 1.55 \mu m$). On a smaller scale the refractive ...ex in the active layer depends upon the carrier density. Typical values reported for the change in index with carrier density are $dN/dn = -2.8 \times 10^{-20} cm^3$ (Manning *et al.*, 1983) and $-1.8 \times 10^{-20} cm^3$ (Westbrook, 1986). This variation causes a shift of the allowed cavity modes towards shorter wavelengths. As temperature increases two effects occur: the band gap decreases so that the peak of the gain spectrum moves to longer wavelengths, and at the same time the refractive index increases so that the wavelength of each individual longitudinal mode also increases. The net effect is again a shift of peak wavelength from one mode to the next at a rate 0.2–0.7 nm/deg C depending on the wavelength of operation (0.85–$1.55 \mu m$). Thus by varying both the laser current and temperature the emission wavelength can be tuned. Typical tuning ranges of ~ 10 nm are possible as the temperature changes from $15°C$ to $40°C$, although this depends upon the characteristics of the individual device (T_o, thermal impedance, threshold). In general, the behaviour of laser spectra is not completely understood; both smooth mode changes and sharply defined mode hops have been reported as the device current is increased, and hysteresis effects have also been seen (Nakamura *et al.*, 1978; Camparo and Volk, 1982).

The discussion so far has considered the performance of an idealised device in that the output in each longitudinal mode is assumed constant. In practice, however, noise within the cavity (i.e. electron and photon density fluctuations) leads to variations with time of the output power in each mode (Ito *et al.*, 1977), known as mode competition noise (see Section 7.3). Under transient conditions, even for a normally single-mode laser (as measured cw) there will usually be periods when the device operates in many modes. For example, when switched on from below threshold, the transient spectrum will begin with many modes and then narrow down to one predominant mode during the first few nanoseconds of operation. Thus there can be considerable spectral broadening when a laser is modulated with short current pulses such that the device goes below threshold between pulses.

For long-distance optical communications using dispersive fibres there is a need for single-longitudinal-mode (SLM) operation of laser diodes. This is difficult to achieve in conventional laser structures as a consequence of the relatively broad spectrum of stimulated emission as compared to the mode spacing. This situation is improved by the use of quantum-well devices, where the stimulated spectra are considerably narrower than for conventional double heterostructures (Burt, 1983); thus $1.55 \mu m$ quantum-well lasers show a tendency towards SLM operation (Tsang, 1984). Another approach to SLM is the use of short cavity lengths (Lee *et al.*, 1982); reducing L to $70 \mu m$ in eqn. 3.13 gives a mode spacing of 4.3 nm. However, even with this value of $\delta\lambda$, mode hopping may still occur under conditions of large-signal modulation, with consequently serious deterioration of available bit rate. A better solution is to modify the laser cavity by the introduction of extra reflecting surfaces; some of the possibilities which may be envisaged (Smith, 1972) are shown in Fig. 3.19.

These configurations all work on the principle of superimposing two or more sets of Fabry–Perot cavity resonances in order to favour certain modes (where the FP modes coincide) at the expense of others (where the FP modes are at different wavelengths). This principle is illustrated in Fig. 3.20 for two sets of FP resonances, such as would be the case for the cavity of Fig. 3.19a. Practical structures are discussed in Chapter 4.

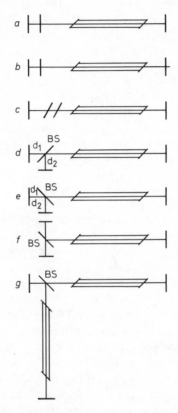

Fig. 3.19 *Compound laser cavities (after Smith, 1972)* © 1972 IEEE

A completely different way of achieving SLM operation is the use of DBR or DFB structures as described in Section 3.5. For the former device, in order to achieve efficient SLM operation, one DBR (length L_2) must have low reflectivity for optimal output coupling and the other DBR must have narrow bandwidth and high reflectivity for SLM single-ended emission. The mode spacing $\delta\lambda$ is given by an equation similar to eqn. 3.13, but with the length L replaced by an effective length L_{eff} which is given to a good approximation by (Stoll, 1979)

$$L_{eff} = L_a + \frac{1}{2}\left[\frac{L_2}{\alpha_g L_2 + 1} + \frac{1}{\alpha_g + (\alpha_g^2 + \kappa^2)^{1/2}} \right] \qquad (3.22)$$

where L_a is the length of the active region, α_g is the loss per unit length in each DBR, and κ is the grating coupling coefficient. In order to ensure SLM operation, the FWHM of the reflectivity of the strongest DBR (since this dominates the cavity bandwidth) should be less than the mode spacing $\delta\lambda$.

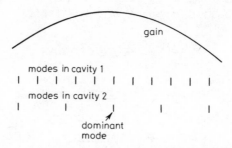

Fig. 3.20 *Operating principle of the double-cavity structure of Fig. 3.19a*
The gain spectrum is superimposed on two sets of modes corresponding to the resonant wavelengths of the two cavities

For the DFB laser the oscillating wavelength is located close to, but just outside, the stop band of the grating (see Fig. 3.14) in order to allow propagation. For the high-gain situation, the resonant modes are separated by $\delta\lambda$ as given by eqn. 3.13 and are located symmetrically on either side of the stop band (note that oscillation will *not* occur actually at the Bragg condition). This means that in principle the two modes closest to the Bragg wavelength have an equal chance of lasing. In order to overcome this potential disadvantage and achieve true SLM operation, Haus and Shank (1976) proposed the introduction of a $\pi/2$ phase shift between the two halves of the DFB grating. This has the effect of creating a pass band in the centre of the grating stop band (Kim and Fonstad, 1979), and thus permitting lasing at the Bragg wavelength. Such a structure has been realised and SLM operation demonstrated by Sekartedjo *et al.* (1984). A significant advantage offered by the DFB (and DBR) lasers is that the lasing wavelength is relatively insensitive to temperature. At $1\cdot55\,\mu\text{m}$, the variation is about $0\cdot1\,\text{nm}/\text{deg C}$, which is roughly one fifth of the value for FP lasers. Another advantage of these devices is their potential for integration with other optical components, e.g. waveguides, modulators, switches etc.

Chapter 4

Physical structures of long-wavelength lasers

There are a wide variety of parameters which need controlling in a single laser design in order to provide a satisfactory communications laser. It needs to emit as much power as possible in the fundamental mode (see Section 3.3) while being operated continuously at room or elevated temperatures, and be able to do so over a lifetime of tens of years while maintaining the required spectral and transient properties. The operating current should be low enough to avoid thermal runaway through self-heating, a requirement which is exacerbated by the high temperature sensivity (low T_o – Section 3.6) of GaInAsP lasers. In many cases the laser is bonded with its heat-generating epi-layers adjacent to a copper or diamond heatsink to provide an effective thermal design.

Much of the early research on long-wavelength lasers used double hetero-structures to confine the carriers and light in the vertical direction (Section 3.3) and a stripe contact to restrict current injection in the lateral direction (Section 3.7 and Fig. 3.16). Such stripes were defined by dielectric isolation (Hsieh and Shen, 1977), proton bombardment (Hsieh et al., 1976), reverse bias junction (Oe et al., 1980) or by Schottky-barrier isolation (Bouley et al., 1981). However, it was difficult to obtain simultaneously CW capability at reasonable temperatures and a stable fundamental mode, due to the low T_o and, probably, Auger recombination. Operation to about 80°C could be achieved but with poor yields. Also 'kinky' light/current characteristics could be caused by mode changes or by the lack of lateral control, allowing shifts in the lasing filament as a result of minor perturbations (see Thompson, 1980).

4.1 Lateral optical confinement

The major problems of simple stripe lasers can be overcome by imposing a lateral refractive-index profile (see Section 3.7), the strength and width of which should dominate profile changes due to the thermal distribution or those induced by injected carriers as well as uncontrolled variations produced by the growth and processing, and yet it should be insufficient to confine higher-order modes. The effective index can be altered by changing the local index of any part of the structure occupied by the optical distribution, namely active or guiding layers. In a direct approach Hsieh (1979) used zinc diffusion to increase the index of the stripe region, and obtained $\sim 8\,\text{mW}$ fundamental mode power in CW lasers with threshold currents down to 70 mA. The structure is difficult to control because of the need to precisely control the zinc-concentration profile.

The active-layer thickness can be varied by using LPE growth over a channelled substrate. Fig. 4.1a illustrates such a structure (Doi *et al.*, 1979), and although the mode was stabilised, the position of the active layer directly on a processed substrate presents a reliability hazard. Takahashi *et al.* (1980) using a channelled substrate and Moriki *et al.* (1980) using a terraced substrate (Fig. 4.1b) employed an intervening InP layer to overcome this hazard while still obtaining a stabilised mode. The evanescent tails of the optical distribution can also be altered by removing the top guiding layer adjacent to the stripe, thus producing a rib-waveguide laser (Kaminow *et al.*, 1979). For better protection the material can be replaced by a second stage of epitaxy to produce a buried rib structure (Nishi *et al.*, 1979; Prince *et al.*, 1980; Fig. 4.1c). Alternatively an inverted version can be produced by growth in a groove in the substrate (Sakuma *et al.*, 1980; Turley *et al.*, 1981; Fig. 4.1d, Sakai *et al.*, 1981).

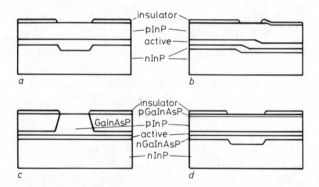

Fig. 4.1 *Structures with lateral optical confinement*
  a Channelled substrate (Doi *et al.*, 1979)
  b Terraced substrate (Moriki *et al.*, 1980)
  c Buried rib structure (Prince *et al.*, 1980)
  d Inverted rib waveguide (Turley *et al.*, 1981)

All these optically confined structures have been, or can be, designed to provide a stable fundamental mode, but suffering loss of carriers by current spreading or lateral diffusion in the active layer. Threshold currents of these types of structures have been reduced to 38 mA with fundamental mode output powers of 30 mW (Horikawa *et al.*, 1984). Potentially a major advantage of these structures is the absence of processes surfaces immediately adjacent to the active region, thus avoiding many reliability hazards (Rosiewicz *et al.*, 1985).

4.2 Lateral optical and carrier confinement

The development of long-wavelength lasers with lateral optical and carrier confinement followed the techniques used for GaAlAs lasers. The simplest structure was the mesa (Fig. 4.2a) in which the lateral regions are removed. This structure is difficult to contact to, has a long exposed p–n junction, and must be very narrow in order to operate only in the fundamental mode. Contacts can be made with a separate pad (Chen *et al.*, 1980; Fig. 4.2b) or by using insulating layers of alumina (Blondeau *et al.*, 1982; Fig. 4.2c) or CVD silica and polyimide (Koren *et al.*, 1983). Threshold currents can be low, 14 mA (Blondeau *et al.*, 1982), and fundamental mode power can be high, to 24 times I_{th} (Koren *et al.*, 1983), but with absence of reliability data the devices will remain of research interest only. The use of dielectric isolating layers does provide some passivation, but crystalline passivating layers are preferred, i.e. the buried heterostructure lasers.

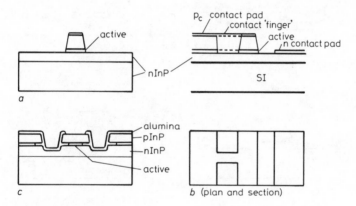

Fig. 4.2 *Types of mesa laser*
a Basic structure
b Mesa with contact pad (Chen *et al.*, 1980)
c Mesa with insulating alumina layer (Blondeau *et al.*, 1982)

The ideal buried heterostructure (BH) laser, illustrated in Fig. 4.3a, consists of a mesa sandwiched laterally between insulating material which prevents current bypassing the active layer or diffusing from it. In order to obtain fundamental-mode operation, most structures are designed so that all higher-order modes are cut off; however, Buus (1981) showed that consideration of threshold conditions leads to a less critical design. Koch *et al.* (1984), in the only direct version of this, used semi-insulating VPE material and obtained threshold currents of 17 mA and output power up to 6 mW.

Earlier work on BH lasers used blocking layers with one or more reverse biased p–n junctions. These could prevent current flowing directly from the contact, but they allowed small currents to bypass the active layer. Hsieh and

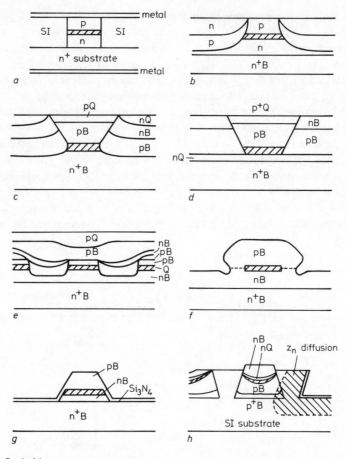

Fig. 4.3 *Buried heterostructure types*
a Ideal buried heterostructure
b Buried heterostructure with *n–p* blocking layers (Hsieh and Shen, 1977)
c Improved BH with *n–p* blocking layers (Kano and Sugiyama, 1979)
d Strip-buried heterostructure (Nelson *et al.*, 1980)
e Double-channel planar buried heterostructure (Mito *et al.*, 1982)
f Mass-transported buried heterostructure (Liau and Walpole, 1982)
g Buried heterostructure grown through a masking layer (Chen *et al.*, 1981)
h Three channel buried cresent (Koren *et al.*, 1984)

Shen (1977) made the first GaInAsP BH lasers using the *n–p–n* structure illustrated in Fig. 4.3*b*. In this a bypass current flowed from the contact layer to the *p* blocking layer, even through the forward biased InP *p–n* junction. Hsieh and Shen's pioneering work demonstrated lasers with threshold currents of 105 mA. Kano *et al.* (1978, 1979) reduced the threshold current to 30 mA for 1·3 μm emission and Nagai *et al.* (1980) demonstrated 25 mA operation for the

1·55 μm version. Mizuishi *et al.* (1980), who emphasised the need for low-current operation, in order to carry out high-temperature accelerated ageing tests, used essentially the same structure (Fig. 4.3c) and obtained lasers with threshold currents of 10 mA which could operate CW up to 95°C.

Plastow *et al.* (1982) emphasised that the bypass current depended on the precise positioning of the blocking layers relative to the mesa layers, and that they could be eliminated by using a multiplicity of junctions so that at least one reverse-biased junction lay above the mesa junction and one below it. Their structure has produced threshold currents of 19 mA. Nakano *et al.* (1982) pointed out that the bypass current could become dominant if the *n–p–n* transistor structure of the blocking layers turned on. This occurs at high currents and limited the lasers of Nakano *et al.* to 40 mW output (pulsed). By increasing the *p*-type 'base' thickness they increased the output power capability to 50 mW. By using a *p–n–p* blocking structure on *p*-type substrates, with the higher-resistivity *p* blocking layer adjacent to the active layer Nakano *et al.* (1981) obtained pulsed output powers of 800 mW per facet from a 4 μm wide mesa. Though this was not in the fundamental mode it does demonstrate very effective blocking.

If the etching for an LOC-type mesa (Section 3.3) is terminated at the lower interface of the active layer, the structure illustrated in Fig. 4.3d can be obtained (Nelson *et al.*, 1980). In this the lateral refractive-index step is smaller than in the standard BH, allowing a larger fundamental-mode profile. Nelson *et al.* (1980) fabricated this strip-loaded BH and obtained over 500 mW pulsed output power.

In the BH structures so far, growth over the mesa is prevented by using a suitable mask; however, Mito *et al.* (1982) showed that LPE growth conditions could be adjusted so that blocking layers could be grown at the sides only of an unprotected mesa, and then the blocking layers and the mesa embedded by the next layer grown in the same LPE stage. They obtained threshold currents of 8·5 mA and 13 mA for 1·3 and 1·55 μm emission, respectively, from these planar buried heterostructures. Further modification to improve overall performance and manufacturability resulted in the double-channel planar BH (DCPBH) illustrated in Fig. 4.3e (Mito *et al.*, 1982). 1·3 μm versions of this have had threshold currents of 10 mA, could emit over 50 mW CW power in the fundamental mode and could operate CW up to 130°C.

Liau and Walpole (1982) developed a novel BH technique by converting an undercut mesa structure such as that in Fig. 4.2c to a true BH by using mass transport wherein InP is transported from sharp external corners to internal corners during a heat treatment (Fig. 4.3f). They obtained threshold currents of 9 mA initially and 5 mA in a mass-transport modified etched-facet version. Renner *et al.* (1984) have obtained threshold currents of 4·6 mA for standard-length lasers, fundamental-mode power of 3·5 mW and CW operation to 120°C.

Another structure requiring single-stage epitaxy can be obtained by growing the desired layers through a window in a suitable masking layer. Chen *et al.*

(1981), using this approach, obtained pulsed threshold current lasers of 45 mA from 4 μm wide stripes (Fig. 4.3g). Significant advances in reproducibility were made when the exposed semiconductor was etched prior to growth. This selective growth in a channel has led to the three-channel buried cresent in Fig. 4.3h (Koren *et al.*, 1984). Growth in the outer channels depresses the otherwise enhanced growth rate in the central channel, thus facilitating layer thickness control. Threshold currents of 21 mA have been obtained for 1·51 μm lasers, but fundamental-mode output powers have been limited to a few milliwatts. Eisenstein *et al.* (1984) have shown that this structure, grown on a semi-insulating substrate, has low parasitic capacitance, as exemplified by its 5·7 GHz 3 dB small-signal bandwidth.

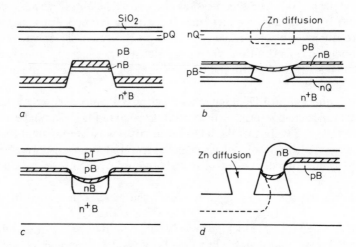

Fig. 4.4 *Buried heterostructures grown on profiled substrates*
a Mesa substrate buried heterostructure (Kishino *et al.*, 1979)
b Double current-confined buried cresent (Murotani *et al.*, 1980)
c Channelled-substrate buried cresent (Devlin *et al.*, 1981)
d Groove transverse junction laser (Chen *et al.*, 1982)

Crystallographically selective nucleation and growth has been exploited in several BH structures. The first such long-wavelength structures were grown over ridges etched in the substrate (Fig. 4.4a). When the mesa was sufficiently high, and had {111A} sidewalls, the initial layers could be discontinuous, and subsequently embedded by a thicker final layer (Kishino *et al.*, 1979). Subsequently Nomura *et al.* (1981) demonstrated operation at 1·3 and 1·58 μm with threshold currents of 20 mA. The ridge of this structure causes a temperature-dependent current crowding effect which improves the temperature sensitivity (see Botez and Connolly, 1980) as confirmed by Tamari and Shtrikman (1982) who measured $T_o \simeq 97$ K on such a structure.

The active-layer stripe of a BH laser can also be obtained by growing on a channelled substrate. In this case the blocking layers may be obtained by etching the channels in previously grown planar blocking layers (e.g. Murotani *et al.*, 1980; Fig. 4.4*b*). By incorporating *p–n* junction isolation as well, Hirano *et al.* (1982) obtained threshold currents of 8 mA for 1·3 μm emission and Oomura *et al.* (1981) obtained output powers of 25 mW. Devlin *et al.* (1981) demonstrated that acceptability low threshold currents of 45 mA could be obtained at 1·54 μm wavelength for a channelled substrate laser which relied on a forward biased *p–n* junction in InP to decrease the bypass current (Fig. 4.4*c*). In another approach Uchiyama *et al.* (1982) used a channel etched through a Be implanted layer, but the threshold currents of 160 mA are probably not representative of the potential.

Chen *et al.* (1982) combined the crystallographic and dielectric masking techniques of selective-area epitaxy by growing in a channel with a Si_3N_4 mask in place on one side only (Fig. 4.4*d*). They also used a microcleaving technique to obtain a cavity length of 38 μm with a threshold current of 3·8 mA, which, with two high-reflectivity dielectric facet coatings, was reduced to 2·9 mA, though the differential efficiency was very low.

Growth in a V-shaped channel with {111B} sidewalls etched through a blocking-layer structure was used by Ishikawa *et al.* (1981) to produce 1·3 μm lasers which could emit 20 mW in the fundamental mode and had a threshold current of 9 mA. Imanaka *et al.* (1984) have increased the output power in the fundamental mode to 60 mW, by using a *p*-type substrate version of the V-channel laser.

4.3 High-power lasers

The ultimate power requirement for communication systems is to maximise the power launched into the fibre without jeopardising the laser's spectral, transient or ageing behaviour. This entails maximising the output power and the laser/ fibre coupling efficiency. In practice, coupling efficiencies of 10–60% are usually achieved for monomode fibres using butt or lensed fibre techniques (Saruwatari and Nawata, 1979; Khoe *et al.*, 1983) though the higher efficiencies are compromised by very tight positioning tolerances. High values are difficult to achieve because of the compromise required between having a strongly confined mode to maintain a stable transverse mode in the laser or a weakly confined laser mode to match the fibre mode. The main limitations to the output power of the laser are listed below:

(i)   The transverse-mode design must prevent the onset of higher-order modes at high currents.
(ii) The blocking layers must not begin to pass current at high laser currents.
(iii) The self-heating of the laser must be minimised by careful attention to the electrical and thermal design. This was discussed for simple stripe lasers by

Steventon *et al.* (1981). Using a similar analysis to that, it can be shown that the thermal limitation is the most significant one. Using typical device parameters of electrical and thermal resistances and threshold-current temperature sensitivity (T_o) of $5\,\Omega$, 50 deg C/W and 60 K, respectively, a laser is likely to be thermally limited at about 500 mA. This can be nearly doubled for the optimistic values of $2.5\,\Omega$, 30 deg C/W and 80 K, or increased to about 3 A for the values $0.5\,\Omega$, 10 deg C/W and 80 K which might be obtained for a laser array. If facet coatings are used to optimise the single-facet output, powers will be thermally limited to a few hundred milliwatts.

(iv) At the high currents possible below the thermal limitation, current densities of 10^5–10^6 A/cm^2 will create a reliability hazard due to electromigration. The output power of GaAlAs lasers is limited by catastrophical optical damage (COD) of the facets to power densities of 2–4 MW/cm^2. Experimentally the COD limit is at least 10 times higher for $1.3\,\mu$m lasers (Temkin *et al.*, 1982) and, considering the larger fundamental-mode distributions possible with long-wavelength lasers, is unlikely to be a problem.

Thus, in practice, long-wavelength high-power lasers need good transverse-mode control, effective blocking layers even at high current, and very careful thermal design. High differential efficiency is also very desirable to minimise direct modulation drive problems. Nelson *et al.* (1980; Fig. 4.3*d*) and Chen *et al.* (1982) have demonstrated devices which achieve the first two requirements up to 500 and 250 mW, respectively, as demonstrated by pulsed operation. Imanaka *et al.* (1984) have achieved 60 mW CW fundamental-mode operation in a $1.3\,\mu$m V-channel laser on a *p*-substrate (at 400 mA). Asano *et al.* (1984) achieved 79 mW (at 700 mA) using a *p*-substrate BH laser, but did not demonstrate fundamental mode operation.

Multi-stripe lasers can emit high power, and if the individual stripes are coupled together they become phase-locked, and retain the single stripe coherence (Fig. 4.5). The considerable interest in high-power phase-locked laser arrays has produced 2.6 W CW from a 40-stripe GaAlAs laser (Scifres *et al.*, 1983). Dutta *et al.* (1984) have obtained up to 100 mW CW from a 10-stripe $1.3\,\mu$m phased array, and higher powers are likely, but so far there has been no satisfactory way of efficiently coupling arrays to narrow-core fibres. Thus the future improvements in coupled power for communications are likely to come from single laser structures with improved thermal designs.

4.4 Single-longitudinal-mode (SLM) lasers

The need for SLM lasers, and the principal ways of obtaining them, were described in Chapter 2 and Section 3.8. The simplest method is to use a cavity which is so short that only one longitudinal mode lies within the gain width. This occurs for photo-pumped platelet lasers (e.g. Stone *et al.*, 1981) and the surface-emitting lasers developed by Iga *et al.* (1983) but neither operate CW at room temperatures. However, very short conventional lasers (50–75 μm) with enhanced facet reflectivities (R_1, $R_2 > 0.9$) can operate under CW conditions predominantly in a single mode (Burrus *et al.*, 1981). Lee *et al.* (1983) have demonstrated that short-cavity BH lasers retain some or possibly all their power in a SLM, by detecting a 274 Mbit/s signal from a laser which was externally filtered to pass solely one mode. Koren *et al.* (1983a) also obtained SLM operation by using reflectivities of 85 and 98·5% on a 37 μm microcleaved laser with 1·3 μm emission.

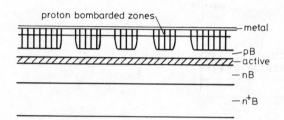

Fig. 4.5 *Schematic cross-section of a possible long-wavelength phase-array laser*

SLM operation can be obtained by the superposition of two or more different Fabry–Perot characteristics using series- or parallel-coupled cavities which may be external or monolithically integrated. The simplest of these is a combination of a laser with a short external cavity tuned to fortify one laser mode (Fig. 4.6*a*). Preston *et al.* (1981) showed that the SLM operation of a 1·3 μm wavelength version with a 200 μm external cavity was established in under 0·5 ns. Subsequently Cameron *et al.* (1982) demonstrated the application of a 1·52 μm version in a 140 Mbit/s system over 102 km. Liou *et al.* (1984) have used a quarter-pitch GRIN rod lens to form the external cavity, and have demonstrated SLM operation under 2 Gbit/s modulation and transmitted over 99 km at 1 Gbit/s. The passive cavity could be an optical waveguide (Fig. 4.6*b*, Stephens *et al.*, 1984) or an integrated optical waveguide (Fig. 4.6*c*; Garmire *et al.*, 1981; Choi and Wang, 1983). The reflectivity required between the two sections could be quite small, and obtainable by incorporating an internal fluctuation in the laser cavity (Fig. 4.6*d*; Choi and Wang, 1982). Coldren *et al.* (1981) achieved such a weak reflection in a two-section laser (Fig. 4.6*e*). The reflectivity, being a function of groove depth, could be strong or weak, but Coldren *et al.* did not achieve stable SLM operation. Ebeling *et al.* (1983), however, achieved SLM operation over a narrow temperature range. However, the complex control circuits and multiple laser contacts required for these will inhibit their use in practical systems.

The two superimposed cavity characteristics can be obtained by coupling two lasers together. Malyon and McDonna (1982) in their 140 Mbit/s 102 km

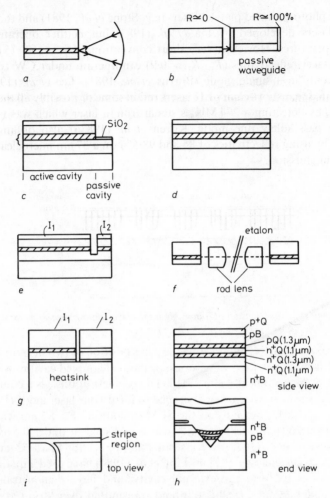

Fig. 4.6 *Multicavity techniques of obtaining single longitudinal-mode operation*
a Exernal cavity in air (Preston *et al.*, 1981)
b External passive waveguide cavity (Stephens *et al.*, 1984)
c Integrated passive waveguide cavity (Garmire *et al.*, 1981)
d Internal-reflection interference laser (Choi and Wang, 1983)
e Two-section laser (Coldren *et al.*, 1981)
f Injection-locked lasers (Malyon and McDonna, 1982)
g Cleaved-coupled-cavity laser (Tsang *et al.*, 1983)
h Double-active layer cresent laser (Tsang *et al.*, 1983*a*)
i Interferometric laser (Fattah and Wang, 1982)

demonstration used a transmitter consisting of a modulated and an unmodulated laser coupled via external optics so that they had only one common mode (Fig. 4.6*f*). Tsang *et al.* (1983) coupled the two lasers together directly on a single heat sink (Fig. 4.6g), and showed that this cleaved-coupled-cavity (C^3) laser could be electrically tuned over 15 nm. 1·3 μm and 1·55 μm versions of this have been used for several systems demonstrations, e.g. 1 Gbit/s over 120 km (Linke *et al.*, 1984), but the complex control circuitry required for stable SLM operation is not attractive.

The possibility of SLM operation by coupling cavities in parallel was demonstrated by Tsang *et al.* (1983a) with a double-active-layer 1·3 μm cresent laser (Fig. 4.6h). The spectral stability may not have been sufficient for system use, but it was shown that the tendency for SLM operation was better for shorter cavities. Fattah and Wang (1982) have also proposed an effectively parallel arrangement (Fig. 4.6i) and showed improved spectral purity for GaAlAs lasers. Neither of these techniques are likely to be of practical importance because of the control problems.

The other approach to spectral purity is to incorporate a diffraction grating. Two versions of the distributed Bragg reflector (DBR) laser described earlier are illustrated in Figs. 4.7a and b. In both cases the grating is in a low-loss waveguide. In one, the integrated twin guide (Kobayashi *et al.*, 1981; Fig. 4.7a), light is coupled from the active layer to the underlying guide and thence to the grating. In the butt-jointed version (Abe *et al.*, 1981; Fig. 4.7b) light is coupled directly to the grating. Mito *et al.* (1983) have made such a butt-jointed laser using DCPBH structure for transverse-mode control and growth over a terrace containing the waveguide and grating (Fig. 4.7c). Threshold currents of the 1·3 μm version were 46 mA at 20°C and SLM operation under CW conditions was demonstrated up to 55°C. The output power was limited to 3 mW, probably by thermal factors. Tohmori *et al.* (1983) added an additional butt-jointed waveguide zone without a grating, and succeeded in tuning the output of a pulsed 1·55 μm version by 0·4 nm, using forward injection current of 4·1 mA in the waveguide zone, while Westbrook *et al.* (1984) obtained 5 nm tuning by controlling the forward current through an active Bragg grating. However, although DBR lasers look useful candidates for SLM communications, they have not been used in any systems demonstrations.

In the distributed feedback laser (DFB) (see Section 3.5), the grating is incorporated in the gain region. Doi *et al.* (1979a) grew the active layer directly on a grating etched on the substrate surface to produce the first long-wavelength DFB laser at 1·1 μm wavelength, but the presence of fabrication-related non-radiative centres within the gain region makes it an impractical structure. The preferred arrangement has a grating in a thin layer adjacent to the active layer (Figs. 3.12c and 4.7d). The problems of creating and then growing over the grating are discussed in Chapter 5.

Progress has been rapid since the first room-temperature demonstration of CW operation at 1·57 μm (Utaka *et al.* 1981). Kitamura *et al.* (1984) have realised a 1·55 μm DC PBH version having $I_{th} = 19$ mA, an output power up to 23 mW and CW capability to 108° C, while 1·3 μm versions have emitted up to 55 mW or 38 mW with or without anti-reflection facet coating, respectively, and

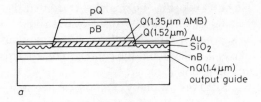

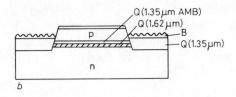

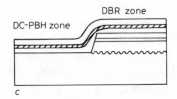

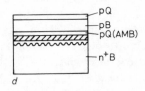

Fig. 4.7 *Distributed Bragg reflector (DBR) lasers and distributed feedback (DFB) lasers*
a Integrated twin-guide DBR laser (Kobayashi *et al.*, 1981)
b Butt-jointed DBR laser (Abe *et al.*, 1981)
c Butt-jointed DBR by growing a DCPBH structure over a DBR terrace (Mito *et al.*, 1983)
d LOC-type DFB laser

have operating CW up to 105°C (Kitamura *et al.*, 1983; Yamaguchi *et al.*, 1984). Also, the mass-transport technique described earlier has been used for 1·53 μm DFB lasers (Liau *et al.*, 1984) with the grating etched through the top surface layers and left uncovered, and in the more conventional approach by Broberg *et al.* (1984).

Although there has been a wide variety of methods of obtaining SLM operation, several are not favoured for communications because they either require very complex control circuitry to maintain SLM operation, e.g. the C^3 laser, or they do not have sufficient mode stability, e.g. short cavity. The use of independently controlled composite cavities has demonstrated their usefulness and acceptably simple control circuitry, but from a systems point of view DFB lasers offer the greatest confidence of SLM operation and can be a direct replacement for Fabry–Perot lasers with the advantage of spectral purity.

Chapter 5

Laser fabrication and performance

5.1 Growth methods

There are four main growth techniques that are used in connection with long-wavelength lasers. These are liquid-phase epitaxy (LPE), vapour-phase epitaxy (VPE), metallorganic vapour-phase epitaxy (MOVPE) and molecular-beam epitaxy (MBE). Of the four, LPE is the most widely used; the others are more expensive and, although they are capable of very large area growth, the problems associated with the material aspects for long-wavelength lasers are still the subject of research. The techniques will be outlined in the following Sections with particular reference to their application to long-wavelength laser materials. In the main, the discussions will focus on the InGaAsP/InP system, the most widely studied for $1 \cdot 0 – 1 \cdot 7 \mu m$ laser sources.

5.1.1 LPE
Growth is normally carried out in a horizonal sliding graphite boat (see Fig. 5.1). In the stationary section there are a number of wells each of which contains the melt constituents required for the successive layers that are required to fabricate the multilayer heterostructure. In addition, there may be melts that are used for *in situ* etching of the substrate just prior to layer deposition to remove thermally damaged surface layers. Before growth starts, the sample, which is held in a recess in the movable slider, is usually covered by an InP slice to reduce thermal erosion. GaAs slices and Sn–In–P in a 'basket' (Antypas, 1980) have also been used for this purpose. The melts are homogenised at the saturation temperature (600–680°C) and then the furnace temperature is lowered. The supersaturation thus generated in the growth solutions provides the driving force for nucleation on the substrates. To accomplish deposition of the desired layered structure, the substrate is moved under each melt in turn, pausing for the correct time in each solution to grow the required layer thickness.

For laser structures of high efficiency the internal non-radiative losses must be low. To achieve this it is essential that the lattice constant of all the layers in

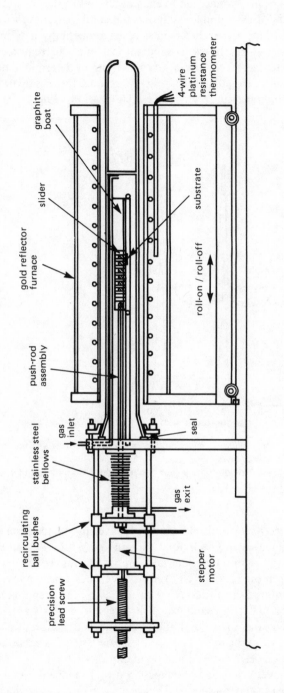

Fig. 5.1 *Cross-section of horizontal sliding boat used for liquid-phase epitaxy (LPE)*

the heterostructure be closely matched to each other and to the substrate. This leads to low interfacial stresses and low interfacial recombination velocities. The correct melt composition of each layer must be experimentally determined to give lattice matching at the growth temperature, and the required lasing wavelength at room temperature. The lattice match can be determined by X-ray rocking curves (Halliwell *et al.*, 1984; Fig. 5.2*a*) and the bandgap can be determined by photoluminescence techniques (Hatch *et al.*, 1982; Fig. 5.2*b*).

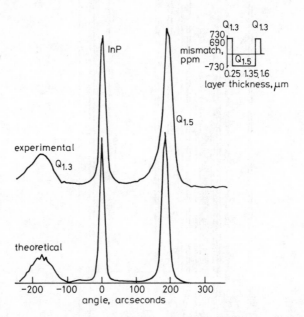

Fig. 5.2 *a Typical X-ray rocking curves showing the angular resolution of diffraction from the substrate and epitaxial layer*
The sign and strength of lattice mismatch can be inferred from the angular separation of the diffraction peaks

Improvements in accuracy and uniformity of wavelength have been demonstrated by using the melt-casting technique (Nelson and White, 1982) illustrated in Fig. 5.3. Here a large quantity of material is prepared, thus reducing weighing errors. This large reservoir of InGaAsP, for instance, is cast into many, e.g. 20, small melts at the required saturation temperature. These are prepared for each composition required in the full sandwich structure, and dopants are added as needed. A variation on this theme that allows the supersaturation of particular layers to be different from the main body of the growth melts is achieved by an *in situ* casting technique (Kawamura and Yamamoto, 1976). This is especially useful for the active layer where good thickness control in a layer of order $0.1\ \mu m$ is crucial. The degree of supercooling, and hence growth rate, can be made lower

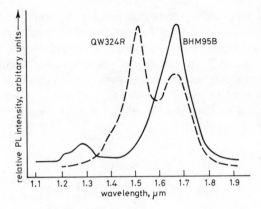

Fig. 5.2 *b Spectrally resolved photoluminescence fro a double-heterostructure laser slice; the stimulus is a 1·06 μm Nd:YAG laser*
The spatial uniformity of the photoluminescence may be viewed with the aid of an IR vidicon camera

for this critical layer. By engineering a growth boat with two sliders the selected melts can be served at the desired intermediate degree of supersaturation.

For wavelengths beyond $\lambda = 1·4\,\mu m$, i.e. as the required As/P ratio in the solid increases, there is a serious problem of dissolution of the InGaAsP layer on bringing the In + InP solution into contact with it (Nagai and Noguchi, 1978). Normally this would be for the growth of the p-cladding layer. The dissolution occurs because of the high As solubility in the solution in equilibrium with the long-wavelength layer.

The use of Sn, rather than the conventional In, solutions to growth n-InP on InGaAs directly has been investigated by Groves and Plonko (1981). Burkhard and Kuphal (1984) have applied this technique to the growth of InGaAsP laser structures. Because the melts contain Sn, the growth layer is inevitably n-type and a p-type substrate is used. There is a relatively high P partial pressure as a result of increased P solubility of InP in Sn (Antypas, 1980). It was found that, even with InGaAs melts, the active layer was InGaAsP as inferred from the lasing wavelength (Groves and Plonko, 1981; Burkhard and Kuphal, 1984), and it was assumed that the phosphorus had been vapour transported from the Sn + InP to the InGaAs melts. The advantage of the Sn solutions is that the growth temperature is very low, 470°C against 590–650°C for In solutions.

5.1.2 VPE

In contrast to LPE, VPE is generally a slow-growth technique, the rate being controlled by gas flows and substrate condition. The apparatus (see Fig. 5.4) is considerably more complex on the gas-handling side and complicated temperature zones have to be established in the furnace. Normally the sample is stationary inside the furnace, although, for reasons discussed later, an externally controlled movement mechanism is sometimes incorporated. The technique can be distinguished into two basic classes depending on the gas system employed. In the first class, the trichloride process, again taking the InGaAsP system as example, $AsCl_3$ and PCl_3 are passed over elemental In or Ga to form metal chlorides, or over binary source wafers such as GaAs and InP. In the second class, the hydride process, metal chlorides are generated by passing HCl gas over hot In or Ga metal and are then combined with cracked hydrides of arsenic (AsH_3) and/or phosphorus (PH_3).

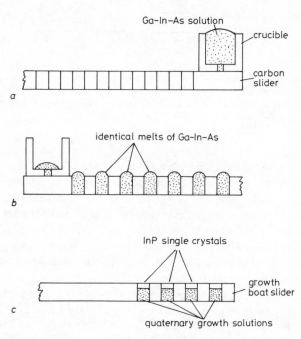

Fig. 5.3 *Melt-casting technique, in which a conventional LPE boat has been extended to allow many melts to be cast from a reservoir, thus improving the accuracy and consistency of individual melts*
a Bake solution to purify and homogenise
b Cast into individual growth melts ($\sim 1\cdot5$ g)
c Combine with INP source material to form two-phase quaternary growth solutions

Two practical advantages that the technique has over LPE are first that the sample size may be more easily increased, and secondly that real-time monitoring of the growth (gravimetric, optical etc.) may be used. Electrobalances have been used to record the change in weight of the sample as the epitaxial layer is deposited (Chatterjee *et al.*, 1982). There is also not the problem of melt back for $\lambda \gtrsim 1\cdot4\,\mu m$ as discussed in Section 5.1.1.

Drawbacks of the technique are the potential for 'hillock' and 'haze' formation, and the problem of interfacial decomposition during the preheat stage.

Because the deposited layer is a function of the gas-stream composition in the vicinity of the sample, it has generally been difficult to achieve sharp interfaces simply by making rapid changes in the inlet gas composition, because of the inevitable delay before the sample's ambient experiences the change. Also the

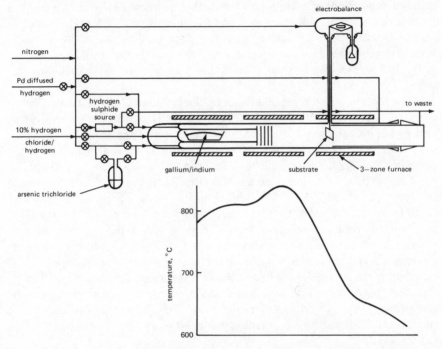

Fig. 5.4 *Schematic of a typical vapour-phase-epitaxy (VPE) reactor used for the growth of InGaAsP*

resultant gas composition may not accurately reflect the inlet changes. It was this fidelity aspect that led to the design of multibarrel VPE reactors (Mizutani *et al.*, 1980). In this scheme the gas compositions required for the target solid compositions are set up and stabilised in separate channels or barrels. To change layer compositions, the substrate is presented to the relevant barrel. Very abrupt changes in doping profile and/or composition have been achieved with this approach. Komeno *et al.* (1983) have studied the electrical properties of the two-dimensional electron gas at the InGaAs/InP heterointerface. Multi-quantum-well (MQW) 1·3 μm InGaAsP/InP lasers grown by VPE have been reported by Yanase *et al.* (1983) and optically pumped InGaAs/InP MQW lasers grown by chloride VPE have been studied by Kodama *et al.* (1984).

However, the VPE technique is mainly being supplanted by metallogranic VPE (MOVPE) as discussed in the next Section. Conventional VPE, having a hot-wall reactor, suffers from AlCl₃ attack of the hot silica glassware, making it difficult to use in the AlGaAs system. This was a major reason for the move to MOVPE.

5.1.3 MOVPE

In this growth technique the group III elements are introduced in the form of metallorganic alkyls, e.g. Ga(CH) or trimethyl gallium, TMGa, as opposed to the chloride or trichloride transport used in conventional VPE, and a cold-wall reactor is used. Only the sample and its holder need be heated in MOVPE, leading to simplifications in reactor design (see Fig. 5.5). Group V hydrides, e.g. AsH_3, PH_3, may again be used, though these can be obtained in metallogranic form.

Growth of InGaAsP is made difficult by the fact that TMIn reacts very easily at room temperature with the hydrides especially PH_3, leading to fumes, formation of additional compounds and rapid depletion of the vapour phase. The premature reaction involves the elimination of methane to give a non-volatile polymer (Didchenko *et al.*, 1960). One proposal to overcome this parasitic reaction has been to use low-pressure MOVPE (Razeghi *et al.*, 1984) as opposed to atmospheric MOVPE. Deposition takes place at reduced pressures ($\sim 100\,\text{mT}$) and the $TMIn/PH_3$ reaction is very much suppressed. The technique has evolved from low-pressure MOCVD Si in which the reduced pressure lowered the dopant vapour pressure above a heavily doped substrate. Without this measure subsequent high-purity layers were unintentionally doped – the autodoping effect. The low-pressure MOVPE technique also has the advantage that the gas composition, and hence the grown-layer composition, can be changed very rapidly leading to very sharp interfaces. Layers that are sufficiently thin and abrupt to show size quantisation effects have been reported (Razeghi *et al.*, 1983).

One solution to the parasitic reaction in atmospheric MOVPE involves the use of adducts (Moss, 1984). These may be pre-prepared or synthesised *in situ*. The pyrolysis of an adduct for the growth of InP, namely $(CH_3)_3$–In–In–$(CH_3)_3$, has been reported (Renz *et al.*, 1979). Adducts for the growth of InP and InGaAs are also reported (Moss and Evans, 1981; Speier *et al.*, 1983).

Growth is achieved by introducing precisely metered amounts of the group III alkyls and the group V hydrides, usually via mass-flow controllers, into a quartz reaction tube. The substrate is placed in this tube on a carbon susceptor which is heated by RF induction. The hot substrate and susceptor have a catalytic effect on the decomposition of the gaseous products and the growth takes place primarily at the hot surface, the substrate acting as a template for the arrangement of the deposited atoms into a lattice-matched epitaxial layer. Typical dopants for *p*-type are diethylzinc, dimethylzinc, dimethylcadmium and (*bis*)cyclopentadienylmagnesium (Nelson and Westbrook, 1984) and for *n*-type H_2Se, SiH_4 and triethyltellurium have been used.

Considerable attention has been paid recently to the problem of reactor design to accelerate the rate at which changes in layer composition can be

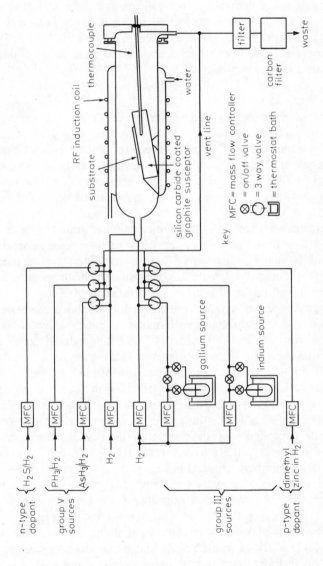

Fig. 5.5 *Layout of a typical atmospheric-pressure metal-organic vapour-phase-epitaxy (MOVPE) reactor*

achieved (see Section 5.1.2). In this respect, laser-holography studies of gas flows in CVD reactors (Gilling, 1982) have been most illuminating. With poor, though up to now commonly used, reactor designs, changes in input-gas compositions are not experienced by the substrate for some considerable time because of gross turbulence upstream of the susceptor. The actual gas compositions eventually arriving at the samples are also considerably distorted because of the turbulence. While this is less of a problem in the AlGaAs system, which is close to lattice matched at all Al compositions used, it is critical in the InGaAsP system because changes in composition must keep the elements in the correct relative concentration to maintain lattice matching.

5.1.4 MBE

The two main material systems that have been studied in this growth technique for long-wavelength laser sources are (a) AlGaInAs/InP and (b) InGaAsP/InP. Both are InP-substrate based. Because of the extra difficulties associated with growing InGaAsP and yet a desire for laser operation at $1 \cdot 5$–$1 \cdot 6 \, \mu m$, devices with InGaAs ($\lambda = 1 \cdot 67 \, \mu m$) active layers have been grown (Temkin et al., 1983), utilising the quantum-size effect to shorten the wavelength by the required amount (Burt, 1983).

In this growth process (see Fig. 5.6), the constituent atoms of the epitaxial layer are supplied to the hot substrate as beams of particles, usually obtained by evaporating the elements in a heated effusion cell. Ultra-high-vacuum conditions are employed and the particles travel without collision to the substrate. Surface kinetics are therefore of primary importance in MBE.

With reference to the two material systems (a) and (b) above, it will be useful to highlight some significant differences in material growth which are peculiar to MBE. At the growth temperature normally used, the group III elements have close to unity sticking coefficients. In other words, almost every group III atom impinging on the hot surface will be incorporated into the growing film. This simplifies the growth of III–III–V and III–III–III–V materials of a particular composition. Only the relative incident fluxes need to be controlled (Temkin et al., 1982). On the other hand, the group V non-unity incorporation coefficients, and the dependence of their surface lifetimes on temperature and surface population, makes the III–V–V and III–III–V–V materials much more difficult to grow at a desired temperature.

Growth of $1 \cdot 5 \, \mu m$ equivalent-wavelength InGaAsP by the standard effusion-cell MBE technique is considerably complicated by the necessity to control the P/As flux ratio. In addition the As cell is relatively rapidly depleted. Recently Panish and Temkin (1984) have overcome these difficulties by using decomposition of AsH_3 and PH_3 as sources of As and P molecules (cf. vapour-phase techniques). This has also been used to advantage in the AlGaInAs system. Prior to this, work had been done in both the GaAs and InP systems (Panish, 1980).

One key advantage of MBE is the ability to clean the surface *in situ* just prior to growth and then determine that it has been cleaned by its reflection electron-diffraction pattern. This glancing-incidence technique shows the surface changing from amorphous to crystalline when successfully cleaned. From the pattern of diffracted spots it is also possible to determine the average atomic ordering of the surface. It can also be used to monitor the film as it grows. Growth of high-quality InGaAs on InP requires that the InP be cleaned of its surface oxide. This is usually achieved by thermal flashing above 500°C for times of order 5 minutes.

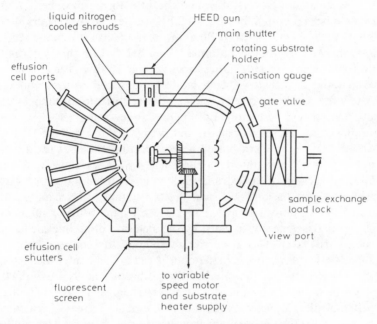

Fig. 5.6 *Plan view of a typical molecular-beam-epitaxy (MBE) chamber*
Elemental beams emerging from the effusion cells impinge on the heated substrate; a selection of *in situ* analysis tools is also shown

Using an As flux it is possible to stabilise the InP (Davies *et al.*, 1980) against non-congruent decomposition by forming a surface InAs layer.

Other advantages follow from the fact that MBE is an ultra-high-vacuum technique. Surface analysis techniques, such as Auger electron spectroscopy, can be included in the growth chamber. Present-day growth kits usually have load-locked entry ports, substrate exchange mechanisms, separate preparation chambers and even linked chambers where dielectric and metal layers can be deposited. These features are all made feasible by virtue of the UHV environment necessary for reproducible growth.

5.2 Fabry–Perot laser fabrication

The majority of laser devices at present still use LPE as the technique by which the active layer is defined. As discussed in Section 4.3, a simple stripe contact rarely suffices when high output power is required.

This problem is alleviated by using one of the buried heterostructures (see Chapter 4). These structures, which include blocking layers, require two stages of epitaxy. Generally both of those stages are LPE, though hybrid growth is now being reported. VPE/LPE has been used for the DCPBH (Yanase *et al.*, 1983a). The BH at 1·65 μm wavelength has been grown by a combination of MBE and then LPE (Kawamura *et al.*, 1982), and the LPE then atmospheric MOVPE approach has been reported for FP (and DFB, see next Section) lasers at 1·5 μm (Westbrook *et al.*, 1983). All-MOVPE at low pressure has also been used in this role (Razeghi *et al.*, 1984). In addition a 1·5 μm BH by LPE then VPE has been reported by Mikami *et al.* (1982). In short, although LPE has been the mainstay of laser-material growth it is being gradually replaced by techniques more suitable to large area homogeneity, and new device structures resulting from the particular features of these growth techniques are evolving.

Returning to the more recent past, and taking for example the DCPBH as representative of the modern buried heterostructure (Mito *et al.*, 1983), the fabrication steps will be outlined. In particular, 1·5 μm wavelength devices will be considered because of the extra layer complexity usually required.

Consider first the basic planar structure. As outlined in Section 5.1.1, dissolution of the 1·5 μm-wavelength active layer by the *p*-InP confining layer when using LPE is generally avoided by the inclusion of an intermediate anti-meltback (AMB) layer. This usually has a composition between $\lambda = 1\cdot1$ and $1\cdot3$ μm. The trade-off is between better optical and carrier confinement, suggesting short wavelengths on the one hand (Westbrook and Nelson, 1984) and active-layer dissolution which needs longer wavelengths. To avoid meltback as the AMB equivalent wavelength is shortened, greater degrees of undercooling must be used with the concomitant increase in growth rates. This makes the desired thin layers difficult to achieve. The use of a quaternary buffer layer below the active layer has also been found to improve the active-layer morphology. This is reflected in lower broad-area lasing threshold current densities (Westbrook *et al.*, 1981).

In the DCPBH structure at 1·5 μm the first growth includes a thin (~0·5 μm) *p*-InP layer above the AMB layer. The first-stage double heterostructure is summarised in Fig. 5.7*a*. Next a masking medium is applied and two parallel channels opened up using conventional photolithography. This mask is then smoothly undercut to generate twin channels which have little evidence of low index crystallographic facets being revealed (Fig. 5.7*b*). Stop-etch planes {111A} similar to the classical BH device will be revealed if the mask is not easily undercut.

Second-stage LPE growth involves *p–n–p* layers. The first two layers are grown in such a way that material is not grown on the central mesa top. This is achieved by controlling the supersaturation of the InP melts and is a manifestation of the so-called 'depletion-mode growth'. This retarded growth is

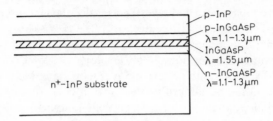

Fig. 5.7 *a The planar double heterostructure grown in the first epitaxy stage of the DCPBH process*
In this starting material the 1.5 μm active layer, shown cross-hatched has a 1.3 μm anti-melt layer above it

Fig. 5.7 *b Twin parallel channels are etched to isolate the active region into a narrow (1–2 μm) strip*

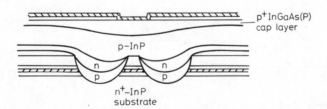

Fig. 5.7 *c Blocking p–n–p layers have now embedded the active-layer strip forming on active optical waveguide*
Current injected through the top contact preferentially flows through the central active layer leading to very low lasing thresholds

commonly seen in LPE growth over relief features, e.g. at the planar shoulders of single channels (Murrell *et al.*, 1983), and is the foundation of many LPE laser technologies (see Chapter 4). The final *p*-InP layer in the DCPBH structure is arranged to grow over the central mesa, thus contacting the *p*-InP layer of the first stage growth. To facilitate the fabrication of low-resistance ohmic contacts, a p^+-InGaAs or p^+-InGaAsP layer is finally deposited on the *p*-InP.

Subsequent steps involve the definition of a contact window in isolating SiO_2 or Si_3N_4 located directly above the buried active layer strip (see Fig. 5.7c). Early devices omitted this current isolation, the view being that the burying layers would block current flow except through the active-layer strip. Although this may have been the case, the blocking-layer p–n junctions produced considerable parallel parasitic capacitance and severely degraded the high-frequency modulation characteristics of the laser (see Section 7.2).

Having defined the p-contact window, the slice is thinned to $\leqslant 100\,\mu m$ to allow cleavage into laser cavities $\sim 200\,\mu m$ long. Alloyed contacts are then deposited, usually AuSn on the n-side and AuZn on the p-side. To improve laser reliability by avoiding an alloyed contact region close to the light-producing active layer, an alternative p-contact technology is sometimes used (Steventon *et al.*, 1981). This involves a shallow Zn diffusion to provide a very high surface p-type doping concentration. Metal contacts applied to this surface are low-resistance tunnelling Schottky barriers. In this respect it was found that the highest Zn-induced hole concentration was obtained at the tenary end of the InGaAsP lattice-matched system. Because low-resistance contacts could then be achieved without alloying and the associated strain and defect loops, more reliable metallisation systems such as Au–Pt–Ti could be used. After metallisation, the slice is cleaved into bars and scribed or sawn into laser chips in the normal way.

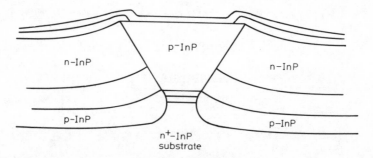

Fig. 5.8 *Cross-section of the classical buried heterostructure*
This again has reverse-biased *p–n* junction blocking; current injected through the wide contact flows through the relatively narrow active layer

There are some significant differences in the fabrication steps of the classical BH (Hirao, 1980). One of the attractions of the design (see Fig. 5.8) is that, because of the reverse mesa shape required prior to the second-stage burying growth, the contact width is large compared to the active-layer width. At $1\cdot3\,\mu m$ wavelength with active layers, $0\cdot1$–$0\cdot15\,\mu m$ thick, clad with InP, the latter should be $1\cdot5$–$2\cdot5\,\mu m$ wide for only the fundamental mode to propagate. For wider dimensions the first-order mode is not cut off and may propagate, leading to undesirable radiation patterns and inefficient fibre coupling.

To achieve consistent laser performance the active layer width needs to be controlled at, say, a tolerance of $0.1\,\mu m$. This implies that the mesa mask and the undercut need to be precisely controlled. Also depth variations of the active layer beneath the top surface translate into width variations. Attempts have been made to overcome this with a combination of etches (Arai *et al.*, 1981).

5.3 DFB laser fabrication

5.3.1 Grating fabrication

DFB/DBR lasers require a diffraction grating to be built into the structure in the vicinity of the optical wave to permit close interaction, and therefore a high degree of wavelength selection. The grating period Λ should be an integral number of half-wavelengths in the material, i.e.

$$\Lambda = \frac{p\lambda}{2N} \tag{5.1}$$

where p is the grating order, λ is the free-space wavelength and N is the equivalent index of the guided medium (see Section 3.5). For first-order gratings (i.e. $p = 1$) the period is typically $190\text{--}240\,nm$ (Mikami *et al.*, 1982). Second-order gratings are sometimes used with a consequent relaxation to $470\,nm$ for $\lambda = 1.5\,\mu m$ operation. However, second-order gratings need to be optimised for maximum first-order content. They also have the feature that some light is dispersed in the direction normal to the grating.

The methods of grating formation generally fall into two types:

(*a*) Holographic lithography
(*b*) Direct-write electron beam lithography

Both of these techniques will now be evaluated [recently first-order gratings have been transferred into resist with X-ray lithography and subsequently into the substrate by ion-beam milling to a final grating amplitude of $0.25\,\mu m$ (Liau *et al.*, 1984)].

(*a*) *Holographic lithography*
This technique is widely used (Akiba *et al.*, 1982; Itaya *et al.*, 1982; Kitamura *et al.*, 1984; Koyama *et al.*, 1984; Olsson *et al.*, 1984). It can be developed to the state of a computer-controlled facility whereby exposure, period etc. can be precisely controlled. Recently five different periods were exposed on the same wafer using a movable opaque mask close to the surface to select parts of the wafer for each exposure (Okuda *et al.*, 1984).

The holographic technique involves the generation of a holographic interference pattern in photoresist on the surface of a semiconductor wafer. Usually a HeCd ($\lambda = 325\,nm$) or a Ar-ion ($\lambda = 458\,nm$) laser is used for first- and second-order gratings, respectively. The optical layout is shown in Fig. 5.9. The laser beam is split and recombined at the water surface at an incident angle θ. Different grating periods Λ are obtained by altering the angle of incidence θ.

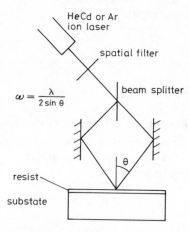

Fig. 5.9 *Optical layout used for holographic exposure of diffraction gratings in photoresist* Light from an Ar-ion laser is split and recombined on the sample at an incident angle θ

The photoresist had previously been applied to the surface by the conventional spin-coating technique followed by post-baking to drive off resist solvents. The different resist sensitivities at the HeCd and Ar-ion wavelengths have an impact on the choice of laser, lower sensitivities requiring longer exposure times and hence tighter constraints on equipment stability. Lateral light and dark fringes of the interfering laser beams expose the resist in a periodic fashion. This technique has been used down to first order at $1\cdot3\,\mu\text{m}$ (Okuda, 1983) where the resist lines are 83 nm wide and the period is 197 nm. Some problems have arisen because of an inability to completely clear the photoresist down to the semiconductor surface. This is because of the large refractive-index step between the photoresist and the semiconductor surface. In addition to the lateral standing wave generated by the two interfering beams, there is a vertical standing wave in the depth of the photoresist caused by reflection at the semiconductor/photoresist interface (Johnson *et al.*, 1968). This is a common problem in photolithography systems where monochromatic light sources have been used to minimise the effects of chromatic dispersion in the optical components. Ripples caused by the vertical standing waves are commonly seen in the resist-edge cross-section. There will always be an intensity null at the resist/semiconductor interface, leading to a residue in the exposed window. Limited over-exposure and/or development can conventionally be used, but is somewhat restricted in the case of high-definition resist gratings. The use of an intermediate anti-reflection layer has also been suggested (Lin, 1983).

After exposure, the photoresist is developed in the normal way. With the submicron dimensions, however, considerable control of temperature, developer strength, particulate count etc. is required.

(b) *Electron-beam lithography*
An alternative method of grating formation uses a scanned focused electron beam to expose electron resist (Turner *et al.*, 1973; Tracy *et al.*, 1974).

Electron-beam lithography has the main advantage of flexibility. By simple software changes, the grating pattern can be directly written, where desired, in any orientation and with a predetermined periodicity. To save electron writing time, gratings need only be written where laser devices are intended (Westbrook *et al.*, 1982). Originally, the strong point behind DFB/DBR lasers was that they dispensed with the need for cleaved-facet mirrors, allowing them to be incorporated in integrated circuits. Electron-beam lithography is ideally suited to this application because associated devices can be delineated in precise location with respect to the laser region.

Phase-shifted gratings have advantages in spectral control of DFB lasers (see Section 5.3.3). In these devices a phase shift of $\Lambda/4$ is introduced in the otherwise period grating. This is very easy to accomplish by electron-beam lithography (Sekartedjo *et al.*, 1984). The device can be thought of as having two separate grating areas spaced by a section whose optical length is an odd number of quarter wavelengths. Holographic lithography has been used to a similar end by predefining SiO_2 masking stripes on the sample before resist application and grating exposure (Koyama *et al.*, 1984). This left uncorrugated stripes on the sample. However, the uncorrugated length, and hence the phase shift between the two sections, could not be controlled as precisely as in electron-beam lithography. Phase shifts have also been obtained holographically using a complex positive and negative photoresist technique (Utaka *et al.*, 1984).

At present only second-order gratings have been reliably fabricated by electron-beam lithography. Even to attain this, special precautions have to be taken to account for proximity effects caused by electrons back-scattered from the substrate. In essence this means that, when a single line is exposed, the adjacent resist is also partially exposed. When a second line is exposed close to the first one, it will need a lower electron dose to be correctly and similarly exposed as the first line. Algorithms can be written to correct for this proximity effect to a certain degree. However, the problem becomes more severe when trying to write the more closely spaced lines of a first-order grating. In order to achieve adequate contrast, higher electron energies have to be used to reduce the relative electron backscatter to workable levels.

Returning to second-order gratings, again the flexibility of electron-beam lithography can be used to optimise the grating. Simple Fourier analysis of the second-order-grating profile (Nelson *et al.*, 1983) shows that the first-order content is maximised for particular combinations of profile and mark/space ratio. In cases where the grating profile is determined by the etches used to transfer the resist pattern to the semiconductor, mark/space ratio can be independently optimised by controlling the electron fluence. Various grating profiles have been demonstrated (Westbrook *et al.*, 1983*a*) in InP with different etch solutions and a thin intermediate SiO_2 mask.

As already mentioned, the grating pattern in resist is transferred to the substrate usually by wet chemical etching (Itaya *et al.*, 1982).

Early workers produced the gratings in the InP substrate. The epitaxial layers, including the active layer, were then grown. More recently, the grating has been formed in a guide layer above the active layer. This has allowed a further refinement. The physical thickness and composition of each layer and the photoluminescence peak can be predetermined. The physical layer information allows the equivalent index N to be more precisely determined. Eqn. 5.1 can be used to match the grating period more closely to the gain peak of the active layer.

5.3.2 Grating amplitude

Large grating amplitudes are desirable to achieve strong coupling between forward- and backward-travelling optical waves via the grating. This leads to low thresholds and increased immunity from unintentional external optical reflections (Yoshikuni *et al.*, 1985). Techniques which use LPE to overgrow the DFB gratings generally suffer from a decreased grating amplitude after growth. For example, Kitamura (1984) formed a first-order grating in the *p*-InGaAsP guiding layer with 240 nm period and 70 nm amplitudes by holographic lithography and chemical etching. After the *p*-type InP cladding-layer growth this amplitude was reduced to 30 nm. Corrugations can be almost preserved for LPE regrowth (Itaya *et al.*, 1984) at reduced temperatures (saturation at 595°C, growth at 578°C). Second-order gratings of 200 nm ampliude decreased to 130 nm after growth.

The reason for the thermal degradation of these fine-geometry structures lies in the so-called 'mass transport' effect. Use has been made of this phenomenon to fabricate novel laser structures (Liau and Walpole, 1982; see Chapter 4). Variation in the chemical potential associated with different surface curvatures is the driving force for thermal deformation of surface-relief features. The equilibrium phosphorus pressures at convex surfaces is higher than at concave surfaces. Vapour-phase transport of phosphorus from convex to concave surfaces occurs and promotes indium transport by surface diffusion. Addition of PH_3 has been reported (Nagai *et al.*, 1983) to dramatically retard this process, thus preventing thermal deformation of DFB gratings. However, there appears to be a discrepancy with other workers, who find PH_3 (Liau and Walpole, 1982; Nelson *et al.*, 1983) or an InP cover wafer (Chen *et al.*, 1982*a*) promoted the mass-transport effect. Nelson *et al.* (1983) found that addition of PH_3 to the H_2 gas stream enhanced thermal deformation, as opposed to H_2 only at the same temperatures. Because phosphorus is believed to play an important role in the mass-transport process, device structures where the grating has been formed in quaternary guide layers instead of the InP substrate are advantageous because they contain less phosphorus. By tailoring the MOVPE process in line with the experimental observations on the effect of PH_3, deformation-free overgrowth of second-order corrugations, 460 nm in period and 160 nm deep, has been achieved (Westbrook *et al.*, 1983).

The same technique of using MOVPE to overgrow DFB gratings without altering their shape and size has been reported by Razeghi *et al.* (1984). MBE has also been used and the gratings remained intact (Asahi, 1983). Before growth, the InGaAsP waveguide surface was thermally cleaned at 550°C for 5 min under As and P (1 : 3) vapour pressures (Davies *et al.*, 1980). Growth of Be-doped InP was then carried out at 480°C.

Also, the mass transport technique described above has been used for 1·53 μm DFB lasers (Broberg *et al.*, 1984).

5.3.3 Techniques for spectral control in DFB lasers
In principle, the conventional DFB laser has two spectral modes or wavelengths of operation with exactly the same required threshold gains. The longitudinal modes are equidistant from the Bragg wavelength, i.e. on either side of and defining the 'stop band'. However, this only holds for completely symmetric homogeneous, infinite structures. Normally this symmetry of period is inter- rupted and the degeneracy removed when the output facet is cleaved, for example. The asymmetry in the required threshold gain caused by reflection at one or more facets or external reflectors has been studied (Streifer *et al.*, 1975; Itaya *et al.*, 1983; Akiba *et al.*, 1983; Chinn, 1973).

The phase of the grating where it is interrupted by the cleaved output mirror, the facet phase, is an important parameter. The importance of this facet phase in determining the lasing threshold and oscillation wavelength has recently been experimentally demonstrated (Matsuoka *et al.*, 1984). Using the precision possible with ion-beam milling to adjust the phase of the facet reflector, they found a cyclic variation in both the threshold current and oscillation wavelength, the metric period of which correspond to $\lambda/2$ within the laser cavity.

Other methods of removing the spectral degeneracy have been discussed. These include partial loading (Kawanishi *et al.*, 1979), use of tapers (Haus and Shank, 1976) and non-uniform stripe (Tada *et al.*, 1984). The most successful to date has been the use of phase-shifted gratings (see earlier references in Section 5.3.1).

Chapter 6

Reliability of long-wavelength lasers

In general, the achievement of reliable long-wavelength lasers has been more straightforward than in the AlGaAs/GaAs case. Problems of defect generation within the active layer and catastrophic failure of the facet mirror at high optical power densities, which plagued short-wavelength lasers, have been largely absent at longer wavelength (Fukuda et al., 1982; Takahei et al., 1983). However, because of increased sensitivity of the lasing threshold to temperature, external process-related problems such as thermal-impedance degradation have become more crucial. In addition, because of the requirements to achieve low operating currents and efficient coupling to monomode optical fibres by control of the radiation pattern, the use of the carrier and optical confining properties of buried heterostructures (see Chapter 4) has become more widespread. Attention has therefore been directed toward the efficacy and long-term reliability of the current-blocking layers which are predominantly of the p–n junction type and include the potential hazard of increasing leakage at regrown interfaces.

6.1 Internal optical degradation

Rapid bulk degradation of InGaAsP/InP laser diodes due to crystal defects is generally not a severe problem. Slide processes associated with misfit dislocations and climb processes associated with dislocation loops are not readily observed in the long-wavelength-material system. The dark-line defect when observed in InGaAsP/InP is in the (110) (Emdo et al., 1982) crystallographic direction as opposed to (100) in AlGaAs/GaAs. Another possible source of defects can be associated with mismatch dislocations. This is more of a problem in InGaAsP/InP because of the relative insensitivity of lattice parameter to Al content in the AlGaAs system (see Section 5.1.1) (Komiya et al., 1983). Local imperfections in the crystal do eventually give rise to dark spots which can be seen in the plan-view electroluminescence image of the active layer. Facet oxidation is ~ 1000 times less than AlGaAs/GaAs lasers (Fukuda, 1983).

6.2 Thermal-impedance degradation

The reliability of the heat-removal process is very dependent on the chip-metallisation, solder and heat-sink technology used. Ohmic contacts are generally alloyed AuGe(Ni) or AuSn for the *n*-side and alloyed AuZn or non-alloyed AuCr and AuPtTi for the *p*-side. Reaction between the *p*-side metallisation and the laser chip (Chin *et al.*, 1984) is generally reduced by the use of the latter metal system with Pt acting as a barrier layer. Chin *et al.* (1984) suggested that Au migration from the *p* contact during device processing and ageing resulted in the formation of dark-spots defects in InGaAsP/InP light-emitting diodes. Metal interdiffusion, especially between the Au contact and In solders close to the heat-producing active layer, can increase thermal impedances.

Intermetallic formation is usually associated with the generation of voids (Van Gurp *et al.*, 1984). Solders in the range In, Sn, PbSn, AuSn and AuGe are commonly used. It was originally used for GaAs lasers because of the need to have a low-stress soft solder. Otherwise defects were easily propagated and the lifetime shortened considerably. Sn whiskers growth from AuSn ($10:90$ $232°C$ eutectic) can cause devices to be suddenly short-circuited (Fukuda *et al.*, 1984). No whisker growth is observed for higher Au content solders ($80:20$, $280°C$ eutectic). However poorer wetting and greater mechanical stress due to the higher-temperature eutectic accompany the use of this solder. Mizuishi *et al.* (1983) studied the use of PbSn ($40:60$ $188°C$ eutectic). Lead inclusion in Sn solder has been reported to suppress Sn whisker growth.

A variety of heat-sink materials such as Si, SiC, type IIA diamond and copper have been used. SiC has a better thermal-expansion match to InP although its thermal resistance is higher than Cu ($2·7$ versus $4·7$ W/cm deg) (Mizuishi *et al.*, 1983).

6.3 Accelerated ageing philosophy

Degradation of laser devices is evidenced by lasing-threshold-current increase which may be accompanied by a decrease in the differential external quantum efficiency.

The physical mechanisms behind these changes are associated with increased leakage through the blocking layers of the burying structure, an increase of non-radiative centres in the lasing region and an increase in thermal resistance.

Because of the complexity of the problem a pragmatic approach of overstressing is generally adoped. Hard screening tests have been developed in an attempt to isolate devices with latent weaknesses. Screening tests usually involve high temperatures and currents in order to locate weak thermal bonds between the laser and its heat sink. Weaknesses in the current blocking may also be revealed by these techniques. This abbreviates the time taken for the initial rapid degradation generally observed though not understood in long-wavelength lasers.

Electroluminescent-mode (EL), i.e. non-lasing, ageing at high temperature ($\geqslant 100°C$) and high current levels has been found useful for selecting long-lived InGaAsP $1.3 \mu m$ buried cresent lasers (Higuchi *et al.*, 1983). These lasers were mounted, active layer uppermost, on BeO heat sinks. Forward voltage reductions at fixed current were associated with lasing threshold increases. The steady-state increase was independent of ageing current in the 100–300 mA range at 100°C.

Blocking $p–n$ junctions in which the n or p surface had been exposed to high temperatures just prior to the regrowth process were found (Hirano *et al.*, 1983) to degrade rapidly in their blocking efficiency during high-temperature EL-mode testing. Their solution was to adjust the n and p doping levels such that the $p–n$ junction was intentionally displaced from the growth interface. Oomura *et al.* (1983) have modelled the degradation of the blocking junction and obtained agreement with experimental data.

Two-step screening tests have also been used. Ikegami *et al.* (1983) use 100 hours, 150 mA, 70°C EL-mode ageing, selecting only those whose lasing threshold at 50°C has increased by less than 10%. This is followed by 5 mW constant output-power ageing for 100 hours at 70°C, and the devices with more than 10% increase in driving current were eliminated. In addition, ageing at 70°C for a further 3000 hours has been proposed as a practical method to pick high-reliability laser diodes for undersea optical transmission systems (Nakano *et al.*, 1984).

High-power ageing tests of $1.3 \mu m$ DCPBH lasers with high-reflectivity ($\sim 90\%$) rear facets (SiO_2 + Au) have shown stable operation for 6000 hours under constant power conditions of 20 and 30 mW from the front facet at 50°C (Murata *et al.*, 1983). Average increases in operating current at 50°C were 1.3%/kilohour and 2.2%/kilohour at 20 and 30 mW, respectively. These degradation rates can be compared to 0.29%/kilohour at 50°C and 0.47%/kilohour at 70°C for operating at constant 5 mW (Nakano *et al.*, 1984).

Chapter 7

Transient phenomena

7.1 Theory

The time-dependent behaviour of semiconductor lasers can best be discussed in terms of rate equations describing the transient development of electron concentration n and photon density S. For a single-mode laser with spatially uniform carrier and photon population, in the absence of carrier-diffusion effects, these equations are:

$$\frac{dn}{dt} = \frac{j}{ed} - \frac{R_{sp}}{\eta_i} - \frac{c\Gamma}{N} gS \tag{7.1}$$

$$\frac{dS}{dt} = \frac{c\Gamma}{N} gS - \frac{S}{\tau_p} + \beta R_{sp} \tag{7.2}$$

where all the symbols have been defined previously, with the exception of t which here denotes time. The first term of the RHS of eqn. 7.1 represents the rate of pumping into the active region, the second term is the spontaneous emission and non-radiative recombination rate, and the third term is the stimulated emission rate. In eqn. 7.2 the three terms on the RHS represent, respectively, stimulated emission, loss (as characterised by the photon lifetime τ_p), and the fraction of spontaneous emission into the lasing mode. The photon lifetime for a Fabry–Perot laser can be identified, in terms of loss coefficients used previously, as follows:

$$\frac{1}{\tau_p} = \frac{c}{N} \left[\Gamma \alpha_{act} + (1 - \Gamma)\alpha_{cl} + \frac{1}{2L} \ln \left(\frac{1}{R_1 R_2} \right) \right] \tag{7.3}$$

In the steady state ($dS/dt = 0$) for the case $\beta \to 0$, the combination of eqns. 7.2 and 7.3 reproduces eqn. 3.16 for the threshold condition. The nominal threshold current density is then give by eqn. 7.1 with $S = 0$ and $dn/dt = 0$; this result reproduces eqn. 3.17 when R_{sp} is identified from eqn. 3.4.

When a step-current pulse is applied to the laser, there is a delay before the light output occurs, the length of this delay depending on the magnitude of the current and on other laser parameters. To describe this delay and the ensuing transient response, it is convenient to think in terms of an equivalent electron lifetime τ_e, defined as

$$\frac{1}{\tau_e} = \frac{d}{dn}\left(\frac{R_{sp}}{\eta_i}\right)_{th} \tag{7.4}$$

where the subscript th means that the value is to be taken at threshold. For example, in the case of bimolecular recombination ($R_{sp} = Bn^2$) with $\eta_i = 1$, then $1/\tau_e = 2Bn_{th}$. In general, when the carrier dependence of the internal efficiency η_i is taken into account, more complicated expressions are obtained for τ_i. Dixon and Joyce (1979) have discussed the solution of eqn. 7.1 for turn-on delay, in terms of the lifetime τ_e for various recombination processes. For the case when the recombination rate is linearly proportional to carrier concentration ($R_{sp} = \eta_i n/\tau_e$), the standard result for time delay t_d at current density j is given by (Konnerth and Lanza, 1964)

$$t_d = \tau_e \ln\left(\frac{j}{j - j_{th}}\right) \tag{7.5}$$

This result does not include the effects of capacitance charging which may also contribute to turn-on delay in lasers (Lee, 1975). The technique of measuring the delay time as a function of applied current is commonly used to determine the electron lifetime via eqn. 7.5 or more complicated relations corresponding to other recombination processes (Dixon and Joyce, 1979; 't Hooft, 1981). When this method is combined with measurements of spontaneous emission, the radiative and Auger recombination coefficients and the leakage current due to electron drift can be estimated (Su et al., 1982; 1982a; Olshansky et al., 1984).

The laser output following the turn-on delay consists of a series of spikes whose amplitude decays with time after the application of a step-current pulse. This is accompanied by the appearance of a damped sawtooth oscillation in the electron concentration. The situation is illustrated in Fig. 7.1, and may be qualitatively understood as a follows. The electron concentration builds steadily during the delay time until it exceeds the value required for threshold. There follows the onset of lasing action and a rapid depletion of carriers due to fast (stimulated) recombination. When there are no longer sufficient carriers to maintain lasing, the stimulated light output ceases and the entire process commences again. However, on each successive cycle the carrier and photon populations commence from somewhat higher values than on the preceding cycle, and hence the periodic behaviour is damped. A small-signal analysis of the behaviour can be made using eqns. 7.1, 7.2 and 7.4, with the result that the angular resonance frequency ω_r and the damping time τ_d are given by

$$\omega_r^2 = \frac{1}{\tau'\tau_p}\left(\frac{j}{j_{th}} - 1\right) \tag{7.6}$$

$$\frac{1}{\tau_d} = \frac{1}{2}\left(\frac{1}{\tau_e} + \tau_p\omega_r^2\right) \tag{7.7}$$

where τ' is defined as

$$\frac{1}{\tau'} = \left(\frac{R_{sp}}{\eta_i g} \frac{dg}{dn} \right)_{th} \tag{7.8}$$

The quantity τ' takes account of possible nonlinearities in the relations between total emission rate, gain and carrier concentration; it is identical with τ_e in eqn. 7.4 only in the case where these relations are linear.

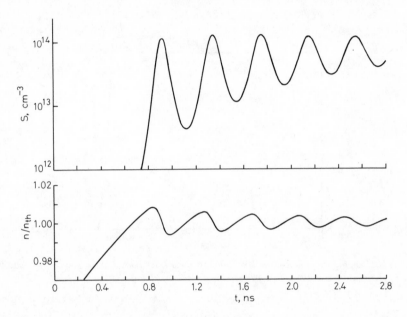

Fig. 7.1 *Relaxation oscillations in photon density S and carrier concentration normalised to threshold* (n/n_{th}) *for a laser*
The curves are calculated for a current pulse of 1.2 times threshold.

The response of a laser to a modulation current may also be analysed with the aid of eqns. 7.1 and 7.2. The result for a small-signal sinusoidal modulation is shown in Fig. 7.2 for typical long-wavelength laser parameters. The response is flat at low frequencies, possesses a peak at a resonance frequency (again given by eqn. 7.6) and then decays rapidly at higher frequencies. The resonance occurs as a result of the existence of the two time constants τ' and τ_p, and is usually in the GHz region. However, at modulation rates above about 100 Mb/s some lasers tend to exhibit relaxation oscillations and self-pulsing effects which may prove a more stringent limit on the modulation rate in practice. As mentioned earlier, however, devices which incorporate a deliberate lateral waveguide structure do not suffer from these problems and exhibit a flat frequency response out beyond 1 GHz. One reason for this is the much lower values of j_{th} for these lasers, so that the overdrive term in eqn. 7.6 can be much larger than for oxide-stripe lasers. In fact, the more advanced devices also exhibit strong damping of the resonance peak seen in Fig. 7.2, as a result of (i) the higher spontaneous emission into the lasing modes, and (ii) the effects of lateral diffusion of carriers. These and other effects will be considered in the next Section.

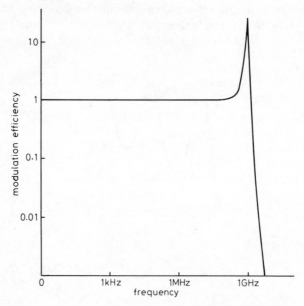

Fig. 7.2 *Calculated modulation transfer function for a laser biased at 10% above threshold, with photon lifetime 1 ps and effective electron lifetime 3 ns*

7.2 Device performance

In this Section we consider a number of additional factors which influence the performance of practical laser structures. These include multimode behaviour, the effects of lateral spatial non-uniformity and device parasitics.

Of importance firstly is the spectrally broad gain in semiconductor laser materials, which results in multimode operation during high-speed modulation of conventional Fabry–Perot lasers. Since a number of longitudinal modes can oscillate, a rate equation must be used to describe the power in each mode, the total output being a sum over all modes. Thus eqns. 7.1 and 7.2 may be rewritten as

$$\frac{dn}{dt} = \frac{j}{ed} - \frac{R_{sp}}{\eta_i} - \frac{c\Gamma}{N} \sum_i g_i S_i \qquad (7.9)$$

$$\frac{dS_i}{dt} = \frac{c}{N} \Gamma g_i S_i - \frac{S_i}{\tau_p} + \beta_i R_{sp} \qquad (7.10)$$

$$g_i = g(\lambda_i) \tag{7.11}$$

Eqns. 7.9, 7.10 and 7.11 describe multimode behaviour where g_i, S_i and β_i are the gain, photon density and spontaneous coupling coefficient of the ith mode, respectively.

This problem has been treated by Adams and Osinski (1982) and Otsuka and Tarucha (1981), who solved numerically multimode rate equations and thereby calculated the time evolution of laser spectra. The main effect on the transient performance of the device of including many longitudinal modes is to couple more strongly the photon and carrier densities, which results in increased damping of transient oscillations (Lau *et al.*, 1983). Since solution of the multimode rate equations is a difficult problem, a common approximation which is used to account for multimode operation is to increase the value of β, the coupling of the spontaneous into the lasing mode (Petermann, 1978). By using this approach a simple wavelength-independent gain function can be used. The effect of increasing β on the relaxation oscillations has been investigated by a number of workers (see for example Boers *et al.*, 1975) who show that increasing β increases the amount of damping during transient oscillations.

Secondly, lateral carrier diffusion has been shown to have an important effect on transient response (Chinone *et al.*, 1978; Furuya *et al.*, 1978); this occurs most strongly in devices which have poor current confinement. To account for this effect the rate equations must be re-written including lateral dependences x (Wilt and Yarviv, 1981):

$$\frac{dn(x, t)}{dt} = \frac{j(x)}{et_a(x)} - \frac{R_{sp}(x)}{\eta_i} + D\frac{d^2 n(x, t)}{dx^2}$$

$$+ \frac{D}{t_a(x)}\frac{dt_a(x)}{dx}\frac{dn(x, t)}{dx} - \frac{c}{N}\Gamma(x)g(x, t)S(x, t) \tag{7.12}$$

$$\frac{dS(x, t)}{dt} = \frac{c}{N}\Gamma(x)g(x, t)S(x, t) - \frac{S(x, t)}{\tau_p} + \beta_{eff}R_{sp}(x) \tag{7.13}$$

The rate eqns. 7.12 and 7.13 include the effects of lateral carrier diffusion and are of a general form suitable for devices with non-planar active layers using $t_a(x)$ as the active thickness. D is the diffusion constant. A single effective value of the spontaneous coupling coefficient β_{eff} is included for simplicity. $j(x)$ allows for the possibility of a lateral current profile.

For devices with lateral optical confinement (a built-in lateral waveguide) but no current confinement, lateral diffusion both damps relaxation oscillations during turn-on and gives rise to 'tails' during turn-off (Henning, 1984). Fig. 7.3*a* shows a diagram of a device with a built-in lateral waveguide but without carrier confinement. Fig. 7.3*b* shows the measured pulse response of the device and Fig. 7.3*c* the response computed from eqns. 7.12 and 7.13. The tail on the optical output arises because of a non-equilibrium carrier distribution during the turn-off transient and results from an enhanced lateral diffusion rate. Lateral diffusion has also been shown to cause similar effects in an LOC structure (Channin *et al.*, 1984).Effects such as these which place limits on the upper modulation bandwidth of a device can be minimised by confining the current to the guide region. Proton isolation has been suggested as a possible method for

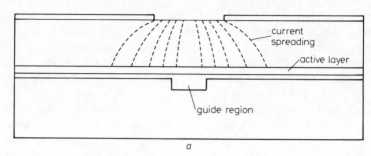

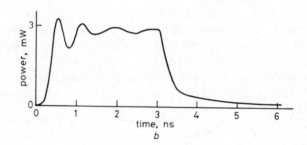

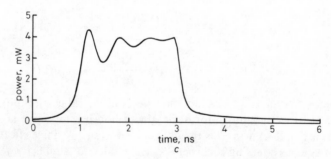

Fig. 7.3 *Device with a built in lateral waveguide but without current confinement (Turley, 1981)*
a Schematic diagram
b Measured pulse response
c Calculated pulse response

lateral current confinement (Henning, 1984). A structure which confines both current and light in the same step is the ridge-waveguide laser. However, because of the continuous active layer under the ridge, this device also suffers from some lateral carrier-diffusion damping. Since the current is confined, the effects are much reduced and high modulation bandwidths have been achieved (Kaminow, 1984).

For devices with both lateral carrier and optical confinement such that the active-layer width is approximately equal to or less than a diffusion length, it has been shown that lateral diffusion both damps relaxation oscillations and reduces the height of the small-signal resonance peak (Furuya *et al.*, 1978). This has been proposed as a possible mechanism for engineering a flat frequency response. Tails due to lateral diffusion are not generally seen in this type of device because the amount of charge injected into regions of low optical intensity is small.

Fig. 7.4*a* and *b* shows the measured pulse response of two different device structures both of which include lateral optical and carrier confinement. Slow rise and fall times are often seen with the rising and falling edges containing components which appear exponential in nature. The modulation rate for these devices would be limited to less than about 1 Gbit/s. This poor modulation

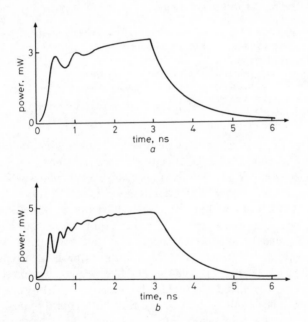

Fig. 7.4 *Measured pulse response of devices with reversed bias current blocking layers*
a 1·3 μm DCPBH
b 1·55 μm CSBH

performance has been noted by a number of workers and has been attributed to parasitic impedances present in the device (Figueroa *et al.*, 1982; Henning, 1984; Tucker *et al.*, 1984). Fig. 7.5 shows a typical circuit model for a laser derived from high-frequency return-loss measurements (various circuit models have been used, see e.g. Figueroa *et al.*, 1982; Henning, 1984). A number of parasitic elements are present, both external (the inductance of the bond wire) and internal (parallel capacitance from current blocking and oxide isolating layers). The main features of both the small and large signal-modulation performance can be explained with the aid of these circuit models. As with lateral diffusion, it has been suggested that the parasitic circuit elements could be useful for engineering a flat frequency response; however, it is doubtful whether a flat phase response could be maintained simultaneously (Bowers and Burrus, 1984).

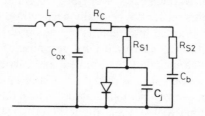

Fig. 7.5 *A typical small-signal equivalent circuit model for a laser*
  L = bondwire inductance
  C_{ox} = a parallel effective capacitance comprising an oxide isolation layer in series with either blocking layers or other junctions
  R_c = contact resistance
  R_{s1} = effective resistance in series with active layer
  R_{s2} = effective resistance in series with blocking layers
  C_b = capacitance of blocking layers
  C_j = diffusion capacitance of device

A further problem which is often observed at high modulation rate is that of the laser output at any particular time being dependent upon the signal during the previous few bit periods. This sort of 'history' or 'memory' effect is known as patterning. It occurs when the modulation rate is such that the carrier density cannot reach equilibrium within a bit period (typically $> 300\,\mathrm{MBit/s}$). In order to minimise this effect it is necessary to achieve short relaxation times (i.e. high resonance frequency) and short damping times (low parasitics).

By optimising device structure, modulation rates have been reported up to $11\,\mathrm{GHz}$ at $0.85\,\mu\mathrm{m}$ (Lau *et al.*, 1984), $18\,\mathrm{GHz}$ at $1.3\,\mu\mathrm{m}$ at room temperature (Olshansky *et al.*, 1986), and $26.5\,\mathrm{GHz}$ at $1.3\,\mu\mathrm{m}$ at $-60°\mathrm{C}$ (Bowers *et al.*, 1985). This has been achieved by using a device with both lateral optical and carrier confinement (thereby minimising lateral diffusion effects), reducing parasitic capacitance elements by reducing the area of current blocking layers, by using a short laser which increases the resonance frequency, and finally by using low-temperature operation which reduces the gain coefficient dg/dn.

To attain even higher modulation frequencies quantum-well lasers appear promising since they exhibit a larger value of gain (dg/dn) coefficient than standard DH structures (Burt, 1983; Uomi *et al.*, 1985). Other methods of achieving high-speed operation include the use of optical bistability (White *et al.*, 1981), Q-switching and saturable absorbers (Ito, 1981).

7.3 Noise

The numbers of photon and electrons in a laser are discrete quantities; similarly the recombination and absorption processes occurring are also discrete and give rise to fluctuations in the photon and electron numbers. These quantum fluctuations are amplified and result in an intrinsic noise source which sets an absolute upper limit to the signal/noise ratio achievable in an optical communication system. A convenient way to make a quantitative analysis of this quantum noise is to incorporate Langevin noise operators F_n, F_S into the rate eqns. 7.1 and 7.2 (McCumber, 1966). These operators have zero mean value and account for the discrete nature of the generation and recombination of carriers, and the emission and absorption of photons, respectively. The spectral densities $\langle |F_n^2| \rangle$, $\langle |F_S^2| \rangle$, $\langle F_n F_S^* \rangle$ of the noise operators may be found in terms of the steady-state photon and electron numbers. It is therefore possible to make a small-signal analysis of quantum noise and thus to derive the photon and electron noise spectra, $\langle |\Delta S(\omega)|^2 \rangle$ and $\langle |\Delta n(\omega)|^2 \rangle$. The resulting spectra are somewhat similar in form to the frequency response of Fig. 7.2 with a characteristic resonance frequency again given by eqn. 7.6.

The amplitude noise of the laser output is usually expressed in terms of the relative intensity noise (RIN) which is proportional to the quantity $\langle |\Delta S(\omega)|^2 \rangle /$ $\bar{S}^2$, where $\bar{S}$ is the equilibrium photon density. For a single-mode laser, theory predicts that the low-frequency RIN increases with injection current below threshold, reaching a maximum for currents close to threshold, and decreasing thereafter as $(j/j_{th} - 1)^{-3}$ (Melchior, 1980). This behaviour has been verified experimentally for several index-guided laser structures in both AlGaAs (Melchior, 1980) and InGaAsP (Mukai and Yamamoto, 1984). For multimode lasers it is found that the total RIN due to all modes is generally lower by orders of magnitude than the RIN exhibited by a single longitudinal mode (Ito *et al.*, 1977; Jackel and Guekos, 1977). The detailed behaviour of the RIN with increasing current for multimode lasers depends critically on the structure via the spontaneous emission coefficient (Arnold and Petermann, 1980). Even for apparently single-mode lasers, it has been shown that the quantum noise in the non-lasing modes can influence the dependence of RIN on injection current (Schimpe, 1983, 1983*a*).

A further complication to the understanding of laser noise arises from the competition between two longitudinal modes. Under certain conditions of current and temperature such that the gain peak lies between two longitudinal modes, the noise in a mode reaches a maximum (Henning, 1983). Fig. 7.6 shows a plot of the noise in a mode measured as a function of laser current. Peaks in the noise current correspond to changes of longitudinal mode, the average spectrum containing two dominant modes of equal power. 100% intensity modulation is observed under these conditions when the gain peak lies halfway between two longitudinal modes and, although the noise of the total intensity increases slightly, it still remains low.

In addition to the amplitude noise discussed above, the associated frequency and phase noise should also be considered, especially when the laser is used as a source for coherent communications systems. In order to take account of these effects, a rate equation for the phase ϕ must be used to complement the rate

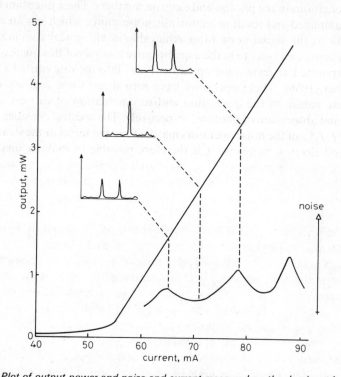

Fig. 7.6 *Plot of output power and noise and current measured on the dominant longitudinal mode as a function of laser bias above threshold*
The insets show the DC (time-averaged) spectra measured at different parts of the characteristic and corresponding to maxima and minima in the noise.

equations for electron and photon densities (eqns 7.1 and 7.2, respectively). The equation for ϕ takes the form (Henry, 1982)

$$\frac{d\phi}{dt} = \left(\frac{c\Gamma g}{N} - \frac{1}{\tau_p}\right)\frac{\alpha}{2} + F_\phi \tag{7.14}$$

where F_ϕ is the Langevin noise operator and α is given by

$$\alpha = \frac{4\pi}{\lambda}\left(\frac{dN}{dn}\right)\bigg/\left(\frac{dg}{dn}\right) \tag{7.15}$$

with λ as the wavelength of the laser radiation. The spectral densities $\langle|F_\phi^2|\rangle$, $\langle F_\phi F_s^*\rangle$ of the noise operators have been derived (Henry, 1982), and thus a small-signal analysis can be used to calculate the phase noise spectrum

$\langle |\Delta\phi(\omega)|^2 \rangle$ (Henry, 1983; Schimpe and Harth, 1983; Spano *et al.*, 1983). Once again, the spectrum exhibits a strong resonance at the characteristic frequency given by eqn. 7.6, in good agreement with measurements on GaAlAs lasers (Daino *et al.*, 1983). The physical interpretation of the phase-noise spectrum arises from the sum of two effects: (i) phase fluctuations of the laser field which result from spontaneous emission, and (ii) electron-number fluctuations which transform to frequency variations via the dependence (dN/dn) of refractive index N on carrier density n. A third effect which can also influence frequency noise at low frequencies (< 1 MHz) is that of current modulation noise which can give rise to temperature fluctuations and thus cause frequency modulation (Tenchio, 1977; Dandridge and Taylor, 1982; Yamamoto, 1983). Inclusion of all three effects can account for the details of observed frequency noise spectra in AlGaAs lasers (Yamamoto *et al.*, 1983).

Phase noise in laser emission is responsible for the finite linewidth of an SLM laser. An expression for the linewidth of an SLM laser. An expression for the linewidth of a Fabry–Perot laser may be found from the full-width half-maximum of the transmission coefficient in eqn. 3.11. However, such an expression would only account for the spontaneous-emission contribution to linewidth, i.e. item (i) above (Schawlow and Townes, 1958). Measurements of the linewidth of SLM AlGaAs lasers showed that the behaviour departs significantly from that predicted by this expression (Fleming and Mooradian, 1981). Henry (1982) showed that inclusion of item (ii), the electron-number fluctuations, in the analysis could give much better agreement with experiment. The general expression for linewidth Δv thus becomes

$$\Delta v = \frac{n_{sp} c^2 hvg}{4\pi N_g^2 P_o L} (1 + \alpha^2)\, \xi \tag{7.16}$$

where P_o is the power output from one facet, n_{sp} is the ratio of spontaneous and stimulated emission (given by the reciprocal of eqn. 3.3), and the other symbols are as defined previously with the exception of ξ which is a normalised end-loss. For the Fabry–Perot laser ξ is given by $-\ln (R)/2$, where R is the facet reflectivity, assumed the same at each end ($\equiv R_1 = R_2$).

The values of the linewidth broadening factor α (as defined in eqn. 7.15) reported in the literature show a rather large spread over the range 2·2–6·6 at room temperature (Henry, 1982; Welford and Mooradian, 1982; Harder *et al.*, 1983; Vahala *et al.*, 1983; Henning and Collins, 1983; Kikuchi *et al.*, 1984; Westbrook, 1986). In general, however, observations of SLM laser linewidth show an inverse dependence on output power P_o, as predicted by eqn. 7.16, with slopes in reasonable agreement with theory. However, in addition a power-independent offset is observed (Welford and Mooradian, 1982a), and has been tentatively attributed to fast thermal fluctuations of the occupancy of states within the bands (Vahala and Yariv, 1983). Measurements of linewidths of 1·3 μm and 1·55 μm multimode lasers have given values much broader than the theory (Olesen *et al.*, 1983), and mode competition noise (Henning, 1983;

Elsässer *et al.*, 1983) and/or mode coupling effects (Elsässer and Gobel, 1984) have been suggested as possible causes for this excess broadening. By contrast, recent results on long-wavelength SLM lasers show narrower linewidths than theory suggests (Kikuchi *et al.*, 1984*a*; Lee *et al.*, 1984). An additional effect is the fine structure which has been seen on the lineshape (Daino *et al.*, 1983; Vahala *et al.*, 1983*a*) which is caused by enhanced fluctuations at the relaxation resonance frequency in agreement with theory (Spano *et al.*, 1983; Henry, 1983; Vahala and Yariv, 1983*a*).

For DFB lasers, the expression (eqn. 7.16) for linewidth still holds true provided that the quantity ξ takes on an appropriate definition. For sufficiently large values of the product of coupling coefficient κ and length $L(\kappa L \gg 1)$ then $\xi \simeq (\pi/\kappa L)^2$ (Adams and Henning, 1985; Kojima and Kyuma, 1984). Fig. 7.7 shows a plot of linewidth versus reciprocal power for a ridge-guide DFB laser (Henning *et al.*, 1984); the device oscillated at $1\cdot47\,\mu m$, 60 nm from the gain peak. The results show the predicted variation with power output, and demonstrate that a linewidth of 15 MHz is possible for an output power of 10 mW. In order to obtain lower values for use in coherent transmission systems, it should be possible to fabricate long devices with high coupling coefficients. However, it must be remembered that as κL increases the slope efficiency of the device falls, and thus for high-κL lasers it may not be possible to achieve high output power due to thermal limitations.

7.4 Spectra under modulation

7.4.1 Multimode lasers

In a direct-detection digital optical communications system employing a laser, bias is generally maintained close to threshold in order to reduce turn-on delay and the modulation current is added on to the steady-state value. Under these conditions (as discussed in Sections 7.1 and 7.2), in response to a step change in current the optical output increases, exhibiting damped relaxation oscillations due to weak coupling between the photon and carrier densities. During this transient period the carrier density also oscillates above and below its steady-state value until after a few nanoseconds (typically < 5) equilibrium is reached. The oscillation in the carrier density is accompanied by a movement of the gain peak, and hence changes of output wavelength (Adams and Osinski, 1982). At low bit rates (where the bit period $\gg$ damping time) the average spectrum under modulation appears very similar to the DC spectrum. However, at high data rates (> 100 Mbit/s) the average spectrum appears broadened and shifted towards higher energies. Moreover competition noise plays an important role in the spectral performance under modulation and causes spectral fluctuations from pulse to pulse (Henning, 1982).

Early experiments using a 1·52 μm laser operating at 320 Mbit/s over 30 km of dispersive fibre indicated that these spectral fluctuations were giving rise to severe system penalties (Frisch and Henning, 1982).

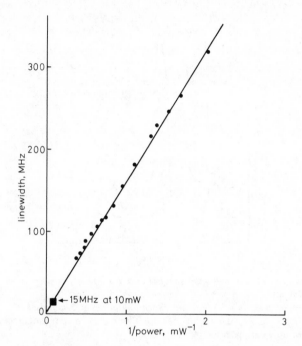

Fig. 7.7 *Linewidth plotted as function of inverse output power measured on a 1·47 μm ridge-guide DFB laser*

In order to assess the effects of partition noise on systems performance, it is necessary to know the spectrum both within a pulse and from pulse to pulse. Prior to the development of a technique, real-time spectra could not be measured and thus much work was devoted to deriving statistics in order to describe the spectral characteristics (Okano *et al.*, 1980; Ogawa, 1982); however, these were generally inferred from averaged spectral measurements (and not real-time data).

In order to measure the true time-resolved spectrum of a semiconductor laser under modulation conditions, a new technique was developed (Henning, 1982). This involved using an array of fibres to temporally disperse the simultaneous signals from each wavelength during a typical pulse. So dispersed, the signals could be recorded and processed to obtain real-time information.

Fig. 7.8 shows the results from such a measurement. The average spectrum under modulation at 320 Mbaud contains five modes, one being dominant. However, the fibre array highlights pulses in which either the optical power is shared between modes or appears in one mode at any of the wavelengths in the average spectrum, i.e. the power can hop between modes. The result given here is for a 1·3 μm laser; however, similar effects were also observed on 1·55 μm lasers.

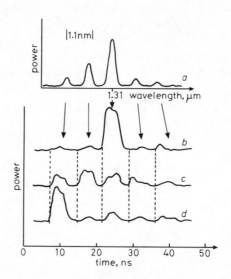

Fig. 7.8 *Upper: plot of time-averaged spectral output of a 1·3 μm laser under modulation at 320 Mbaud*
Lower: output from the fibre array for three different 3 ns pulses
Five fibres were used to gather the output from the five dominant modes; the time slot for the arrival of each mode is indicated

At 1·55 μm the dispersion of a monomode optical fibre is 15 ps/nm/km. From the measurements of true time-resolved spectra it was found that there was a finite probability of the laser emitting at full power at either extreme of the average spectrum in successive pulses. For such a spectral hop of 6 nm an arrival time difference of 3 ns is found for successive pulses over 30 km of dispersive fibre. Thus the power from one bit can appear at the wrong time at the receiver, which produces errors in a digital system (see, for example Ogawa, 1982).

This type of spectral behaviour, observed on a number of Fabry–Perot lasers emitting at 1·55 μm, coupled with fibre dispersion, limits the available (transmission distance) × (bit rate) product. Thus in order to make full use of the lower loss available at 1·55 μm a single longitudinal mode laser is essential for a high-performance system.

As already discussed (see Sections 3.8 and 4.4), by reducing the length of a laser and thereby increasing the longitudinal mode spacing, the average spectrum appears more single-moded. Devices of different length were examined under modulation and their spectral performance is shown in Fig. 7.9.

By measuring the true time-resolved spectrum in the centre of a 3 ns pulse, the RMS spectral width and mean wavelength position were calculated for three devices of different length (Henning and Frisch, 1983). Repeating this measurement on many pulses ($\sim$1000) the probability density functions shown in Fig. 7.9 were generated. The spread in mean wavelength for all three devices is similar, extending some 1·5 nm either side of the most probable wavelength. However, a tendency towards single-mode operation for the shorter (100 μm long) device is seen from the narrowing of the distribution. Calculations based on these results suggested that single-longitudinal-mode operation could be realised for a 70 μm long device with a reflection-coated rear facet. This has been demonstrated by Lee *et al.* (1982).

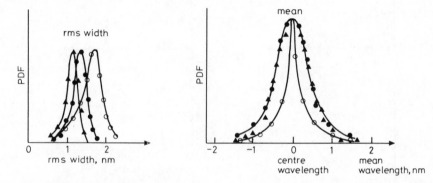

Fig. 7.9 *Spectral statistics after about 1000 events obtained on three 1·5 μm channel substrate buried-cresent lasers of different length*
A tendency towards single-mode operation is seen as a narrowing of the spread in mean wavelength
○ 1316 100 μm long
● 1301 210 μm long
▲ 1309 280 μm long

The influence of external reflections on the spectral performance of a laser is shown in Fig. 7.10. The results given are for a 1·3 μm device in a packing comprising a monomode fibre butt-coupled to the front facet and an angled photodetector for rear-facet power monitoring. The results run in a clockwise direction with increasing temperature. Initially the device shows weak mode selection into mode 4 at 20·4°C; a 0·5 deg C increase in temperature causes much stronger mode selection. Further increases in temperature cause the selection to move towards mode 1. The effect was attributed to power being reflected back into the rear facet from the monitor photodiode.

Similar influence on the spectral output of lasers has been observed for feedback from fibre joints etc. (Dandridge and Miles, 1981; Ikushima and Maeda, 1979). This variability in spectral output can cause penalties in an optical system. For example, if feedback from a fibre joint caused large spectral hops from one pulse to the next, errors may be incurred. Also with mode-selection effects such as shown in Fig. 7.10, a system may exhibit extreme temperature sensitivity with satisfactory operation being maintained only over certain temperature windows where the laser may be almost single-moded.

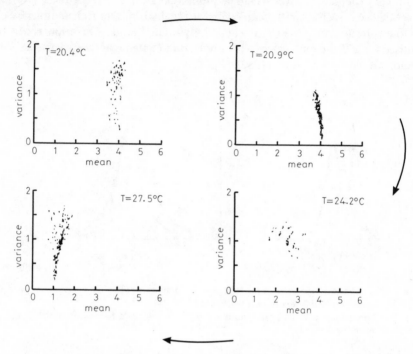

Fig. 7.10 *Plot of variance against mean spectral position for 5000 events each at four different temperatures*
Low values of variance correspond to a tendency towards single-mode operation at one particular mode number; larger values of variance correspond to multimode operation (note that wavelength decreases with increasing mode number)
Packaged 1·3 μm laser

7.4.2 Single-mode lasers

In order to avoid dispersion penalties due to multimode operation, it is necessary to use a laser which oscillates in a single longitudinal mode (SLM) under modulation conditions. Much effort has been devoted to fabricating SLM lasers; however, these also exhibit spectral variations under transient conditions.

As described previously, under high-speed modulation the optical output of the laser exhibits damped relaxation oscillations. During this period the carrier density also oscillates and, since the refractive index within the active layer is a linear function of the free-carrier density, this causes a change of operating wavelength or 'chirp' (Wright and Nelson, 1977, Linke, 1984).

Fig. 7.11 shows the measured optical ouput from a ridge-guide DFB laser in response to a 2 ns pulse; the wavelength chirp is also shown. The effect is most noticeable during the initial turn-on period when the carrier density overshoots its steady-state value before the optical output builds up. The measurements are taken with the device biased just below threshold, whereas biasing above threshold reduces the chirp. Penalties arise because the power in the initial part of the pulse moves into the previous bit period after a long length of fibre; similarily the power at turn off moves into the following bit period (Frisch and Henning, 1984).

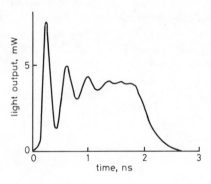

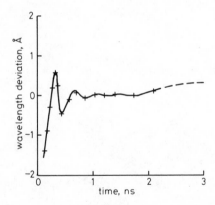

Fig. 7.11 *Light output and mean wavelength position measured as a function of time during a 2 ns pulse*
The device was a ridge-guide DFB laser emitting at 1·47 μm

The amount of chirp depends upon how much the carrier density overshoots its equilibrium value, which is dependent upon a number of factors including the damping in the device, and any parasitics in the laser. In order to reduce the chirp the current pulse could be filtered to reduce the high-frequency content, and hence the carrier period becomes significantly reduced. To a certain extent this could be achieved by design in a device containing significant parasitic impedance through the use of reverse-biased blocking junctions (see Section 7.2); however, variations from device to device would probably be observed.

An alternative approach is to raise the resonant frequency of the device such that the power contained during the first period of transient oscillation (when the chirp is largest) is a small fraction of the total power within the bit period.

The effects of chirp on systems performance are dependent on the device under test. 8 dB of penalty has been seen using a DFB laser at 2 Gbit/s over 71 km, 4 dB of this being attributed to chirp and 4 dB to extinction-ratio penalties (Linke *et al.*, 1984*a*). However, in other experiments somewhat lower penalties have been reported, namely 1 dB at 1·8 Gbit/s over 65 km (Walker *et al.*, 1984). Although the magnitude of the penalty due to chirp is not yet full quantified, it is clear that for direct modulation systems with length–bit-rate products greater than about 100 Gbit km/s, laser chirp is a serious problem which needs further study.

Chapter 8

Conclusions and future possibilities

Many of the future developments which influence laser structures and performance have already been described, e.g. quantum wells, high-output-power
lasers, use of improved epitaxial techniques and narrow-linewidth lasers.
Further developments include optical amplifiers, monolithic integration, very-
high-speed and/or wavelength-multiplexed systems, high-peak-power transmission and ultra-long-wavelength (2–5 μm) transmission. These developments
might be used to improve high-performance point-to-point links, or be used for
quite new types of optical networks.

8.1 Optical amplifiers

Semiconductor optical amplifiers consist essentially of a semiconductor laser in
which an optical input signal is coupled to one facet and the amplified output
signal is obtained from the other facet. Any type of laser structure could be used.
There are three types of amplifier:

(i) *The Fabry–Perot amplifier* (*FPA*) consist of a conventional laser operated
just below its threshold condition. Its spectral characteristics will have the usual
Fabry–Perot mode resonances and the device is thus extremely sensitive to the
wavelength of the input signal. Whilst gains of 25–30 dB can be obtained in
1·5 μm FPAs (Weslake and O'Mahony, 1985), the bandwidth is limited to the
order of 10 GHz. In addition, at relatively modest input levels, the depletion of
carriers (caused by stimulated emission which produces the optical gain) leads
to output power saturation. The saturated output power (defined as the power
at which the gain has decreased by 3 dB) measured on resonance is typically in
the range − 25 dBm to − 12 dBm, depending on the gain of the FPA.

(ii) *The travelling-wave amplifier* (*TWA*) consists of a gain region without a
resonant cavity, e.g. a laser with perfect anti-reflection facet coatings. When the
injection current is high enough the net gain will amplify any injected optical
signal and the spontaneous emission. The latter usually needs to be filtered. By

comparison with the FPA, the TWA has advantages of a wider bandwidth, higher saturated output power, and reduced sensitivity to polarisation of the input signal. The main problem is that of achieving sufficiently low values of facet reflectivity. Recent results on a 1·5 µm TWA have achieved a reflectivity of 0·04%, 3 dB bandwidth of 70 nm (> 9 THz), and a saturated output power of + 7 dBm for 20 dB signal gain (Saitoh and Mukai, 1986). Comparable results for a TWA with 0·08% reflectivity have yielded a maximum gain of 35 dB and a fibre-to-fibre gain of 18·5 dB (O'Mahony et al., 1986; O'Mahony, 1987); the gain difference between orthogonal signal polarisations was measured as 2·5 dB.

(iii) *An injection-locked amplifier (ILA)* is a conventional laser operated above its threshold condition, but driven by an input from another laser. A 1·5 µm ILA driven by $\frac{1}{4}$ to $\frac{1}{2}$ µW from a 1·523 µm HeNe laser emitted 750 µW of power with a linewidth less than 1·5 MHz (Wyatt et al., 1982). However, the locking bandwidth (i.e. the maximum difference between the input frequency and the amplifier resonance frequency that still ensures locking) is rather narrow (around 2 GHz for gains of order 25 dB (Brown, 1986)), and thus extreme stability of the source is required. ILAs may find applications in amplifying phase- or frequency-modulated signals in coherent transmission systems.

There has been considerable interest in optical amplifiers because of their potential applications particularly for optical communications (see Simon (1983), Mukai et al. (1983), or O'Mahony (1987)). They have the following applications:

(i) *Post-amplifier*: to amplify the output of a low-power source, e.g. a 1·523 µm HeNe laser. They are not likely to be used to boost semiconductor lasers because of saturation effects, but they could be useful to compensate for losses due to other components or due to power splitting, particularly when included in semiconductor integrated optical circuits. Post-amplifiers are also of use as modulators as demonstrated by Hodgkinson et al. (1982) in their investigation of homodyne receivers at 1·52 µm.

(ii) *Pre-amplifiers*: to increase the optical signal prior to detection. Simon (1983) pointed out that this would be of greatest interest in direct detection systems, particularly at high data rates, but of little interest for coherent systems where receiver sensitivities are close to the quantum limit. A potential problem for applications as a pre-amplifier is that of amplified spontaneous emission (ASE) from the amplifier, which may impair the receiver sensitivity. However, the use of a narrow-bandwidth optical filter can reduce the effect of ASE and permit very good sensitivity to be achieved (O'Mahony et al., 1986).

(iii) *Repeater*: to amplify a signal attenuated by transmission losses. The considerable simplicity of such a repeater is attractive for undersea applications, particularly if it avoids the complexity of coherent receivers and transmitters. Repeater separations would, however, be smaller than for receiver/transmitter combinations. Smith and Malyon (1982) demonstrated a system using a 1·51 µm amplifier in repeater mode. More recently, systems experiments have proved the feasibility of using two amplifiers in series; a total repeater gain of 26 dB in a 140 Mb/s direct detection scheme has been demonstrated (Marshall *et al.*, 1986). Cascading amplifiers in a system can lead to an additional problem if ASE builds up by amplification in successive repeaters. This problem may be avoided by the use of narrow-bandwidth optical filters.

(iv) *Switch*: where the excellent isolating characteristic in the 'off' state combined with the optical gain in the 'on' state makes the amplifier attractive for use in multiplexing or routing applications. Isolation of 82 dB in the 'off' state and 6 dB gain in the 'on' state has been reported for a $1.3\,\mu$m switch (Kataoka and Ikeda, 1984), and the applications have been demonstrated in optical time switching (Matsunaga and Ikeda, 1985) and in routing (O'Mahony, 1987).

(v) *Optical bistable elements*: in which the output has two stable states for a given input, have recently been analysed and demonstrated using amplifiers at $1.3\,\mu$m (Sharfin and Dagenais, 1985) and $1.5\,\mu$m (Adams *et al.*, 1985). Such bistability could eventually lead to optical logic with switching powers of the order of microwatts and cycle times of a few nanoseconds (Westlake *et al.*, 1986).

Applications in logic gates (Sharfin and Dagenais, 1986) and simple regenerators (Webb, 1986) have already been demonstrated.

These applications, and the ready availability of suitable devices, are leading to an intensive investigation of optical amplifiers. The main problems for most applications are: (i) to obtain a high fibre/amplifier coupling efficiency, (ii) to avoid self-heating effects at the higher current densities that are often used for amplifiers and (iii) to develop perfect anti-reflection coatings for TWAs. The reduced reflectivity coatings which are available are attractive for FPA amplifiers.

8.2 Monolithic integration

Semiconductor integrated optics is currently at the stage where various discrete 'building blocks' have been demonstrated, including waveguides, curved guides, modulators, switches, combiners etc. The major recent advances have been the reduction of switching voltages and propagation losses in the waveguide devices, to acceptable levels (see review by Ritchie and Steventon, 1984). Though much of this work used the GaAlAs system, progress in the GaInAsP system is accelerating rapidly. The limitations, due to chirp, patterning and the resonance frequency in lasers, can be removed by using an integrated modulator. This will benefit high-bit-rate direct detection system, as well as coherent systems.

A loss modulator has been monolithically integrated with a $1.3\,\mu$m DFB laser by Yamaguchi *et al.* (1985); the resultant linewidth under modulation was as narrow as 500 MHz, as compared with a value of 2–3 GHz observed when the DFB laser was directly modulated. More recently, a 1·55 μm DFB laser has been integrated with a GaInAs/AlInAs multiple quantum-well modulator (Kawamura *et al.*, 1986). The use of an integrated passive cavity waveguide to narrow the linewidth of a DFB laser has also been demonstrated (Fujita *et al.*, 1985): the narrowest linewidth reported was 900 kHz at 6 mW output power. Such an integrated cavity can also be used to tune the wavelength of the laser by means of direct electron injection into the waveguide cavity (Sakano *et al.*, 1986). The injected carriers change the refractive index of the cavity and are thus used to control the phase of the optical feedback. Wavelength-multiplexed transmitters are also desirable, and Okuda *et al.* (1984) have already fabricated five DFB lasers on a single chip, operating at different wavelengths. Their output could be combined at a single output guide using suitable waveguide techniques. Similar technology could also be used to combine a laser and detector for two-way operation on a single fibre.

In a separate approach, called optoelectronic integrated circuits (OEIC), electronic components are being developed in materials compatible with lasers and detectors (reviewed by Ritchie and Steventon (1984) and Forrest, (1985)). These OEICs could then consist of monolithic integrated transmitters, arrays of integrated transmitters, complete repeaters or regenerators and, eventually, monolithic chips incorporating receivers, transmitters and digital capability. Current developments include work on GaInAs/InP electronics (Leheny *et al.*, 1980), integratable lasers grown in recesses for improved planarity (Sanada *et al.*, 1984), and using DFB or etched-facet techniques to remove the constraints of designing a circuit between the laser facets (Matsueda *et al.*, 1983). Recently a 1·3 μm DFB laser has been successfully integrated with metal–insulator–semiconductor field-effect transistors and a monitor photodiode (Kasahara *et al.*, 1986); small-signal modulation up to 4 GHz was achieved. Other integration possibilities include very rapid pulse generation from integrated mode-locked transmitters, and eventually optical logic.

8.3 System developments

Though the proven capabilities of optical-fibre systems are impressive, there are further developments which could exploit further the inherent fibre bandwidth. Garrett (1983) in his investigation of the fundamental limits of such systems has suggested that 10–15 dB improvement may be achieved by using different modulation formats. Existing digital systems generally use pulse-code modulation (PCM), but more information can be transmitted with the same energy using, for example, pulse-position modulation (PPM). In PPM a pulse is transmitted in any one of N time slots within a time frame. The full benefit is realised only if the mean transmitted optical power is comparable with that for conventional PCM coding, operating at the same information rate. This requires much higher peak power levels and short pulses. Exploitation of PPM thus needs long-wavelength lasers emitting high-power short pulses 'on demand', i.e. in any desired time slot. Alternatively the output of a regularly pulsing laser can be optically delayed. Progress on high-power picosecond pules obtained both regularly and on demand is reviewed by Ritchie and Steventon (1984), but there has been no published experimental work on PPM systems.

So far all the systems described use silica-based fibres. These now have losses limited by the fundamental IR absorption tail and Rayleigh scattering. Goodman (1978) suggested that other materials with longer-wavelength IR tails could provide a loss minimum of 10^{-2}–10^{-4} dB/km at wavelength of 2·5–5 μm. To date, short lengths of fibre have provided losses of a few dB/km, but progress is currently rapid, and significant improvements may be expected; however, severe problems of fibre strength and glass stability also need to be solved. Such systems would require lasers in the 2–5 μm region. Kobayashi *et al.* (1980) have reported some work in III/V materials and lasers, and other systems are also potentially viable e.g. IV/VI, II/VI and ternary chalcopyrites. Progress of lasers for spectroscopic purposes, e.g. 4 μm emitting PbEuSeTe lasers (Partin, 1984), would be expected to benefit the future communications needs. The reduced losses also imply a reduced stimulated Brillouin scattering threshold of about 50 μW, which will detract from the benefits of systems which require spectrally pure lasers.

The limitation on the maximum bit rate for a fibre system is dictated by dispersion, because the linewidth of a pulse is Fourier-transform-limited according to $\Delta t \Delta v \simeq 0\cdot 36$. The dispersion can be minimised by operating at the wavelength where the first-order dispersion is zero (approximately 1·3 μm), but then the higher-order terms limit the bandwidth (Kapron, 1977); e.g. for a 20 km length the bit rate is limited to 130 Gbit/s (Unger, 1977). This maximum bandwidth falls very quickly as the wavelength departs from the first-order dispersion zero. Hasegawa and Tappert (1973) proposed balancing the dispersive effect with a nonlinear change in the refractive index, the Kerr effect. They predicted this could yield 1 Tbit/s bandwidth for 10 W pulses. Mollenauer *et al.* (1980) verified this soliton phenomena for the anomalous dispersion region, i.e. for $\lambda > 1\cdot 3$ μm where the increasing waveguide dispersion causes net dispersion to increase with wavelength. However Blow and Doran (1983) showed that pulse interactions reduce the bandwidth by one order in the anomalous dispersion region, and at the zero point (Blow *et al.*, 1983) and the normal dispersion region (Nelson *et al.*, 1983). These nonlinear effects occur in monomode fibres with peak pulse powers of > 100 mW, which, although more than an order larger than currently used, could be attainable with short-pulse semiconductor lasers.

8.4 Conclusions

The explosive development of optical communications from the 0·85 μm region to the 1·3 and 1·5 μm wavelength has resulted in a very rapid progression from early simple long-wavelength lasers to sophisticated linewidth-narrowed single-frequency devices. Optical communications has undoubtedly provided the driving force for this, but even though long-wavelength systems are now available commercially, further laser developments are now required for coherent communication systems and for the rapidly expanding sensor market, as well as for the many other improvements which will give rise to higher performance or lower costs. The substantial amount of fundamental knowledge which exists for long-wavelength lasers will provide a good foundation for the newer developments, but the considerable additional research and development which will be required will maintain the excitement for scientists and engineers for many more years.

References

ABE, Y., KISHINO, K., SUEMATSU, Y., and ARAI, S. (1981): *Electron. Lett.*, **17**, pp. 945–947

ADAMS, A. R., ASADA, M., SUEMATSU, Y., and SHIGEHISA, A. (1980): *Japan J. Appl. Phys.*, **19**, L621-L624

ADAMS, M. J. (1981): 'An introduction to optical waveguides' (Wiley) chap. 8, pp. 278–369

ADAMS, M. J., and OSINSKI, M. (1982): *IEE Proc.* Pt. I, **129**, pp. 271–274

ADAMS, M. J., and HENNING, I. D. (1985): *IEE Proc.* Pt. J, **132**, pp. 136–139

ADAMS, M. J., WESTLAKE, H. J., O'MAHONY, M. J., and HENNING, I. D. (1985): *IEEE J. Quantum Electron.*, **QE-21**, pp. 1498–1504

AIKI, K., NAKAMURA, M., KURODA, T., and UMEDA, J. (1977): *Appl. Phys. Lett.*, **30**, pp. 649–651

AINSLIE, B. J., BEALES, K. J., DAY, C. R., and RUSH, J. D. (1981): *IEEE J. Quantum Electron*, **QE-17**, pp. 854–857

AINSLIE, B. J., BEALES, K. J., COOPER, D. M., DAY, C. R., and RUSH, J. D. (1982): *Electron. Lett.*, **18**, pp. 842–844

AKIBA, S., UTAKA, K., SAKAI, K., and MATSUSHIMA, Y. (1982): *Electron Lett.*, **18**, pp. 77–78

AKIBA, S., UTAKA, K., SAKAI, K., and MATSUSHIMA, Y. (1983): *IEEE J. Quantum Electron.*, **QE-19**, pp. 1052–6

AKIBA, S., UTAKA, K., SAKAI, K., and MATSUSHIMA, Y. (1983): 4th Int. Conference on Integrated Optics and Optical Fiber Communications, Tokyo, 27–30 June, pp. 156–157

ALEXANDER, F. B., BIRD, V. R., CARPENTER, D. R., MANLEY, G. W., McDERMOTT, P. S., PELOKE, J. R., QUINN, H. F., RILEY, R. J., and YETTER, L. R. (1964): *Appl. Phys. Lett.*, **4**, pp. 13–15

ALFEROV, Zh. I., and KAZARINOV, R. F. (1963): Author's Certificate 181737, Claim No 950840, as cited in Alferov *et al.* (1970).

ALFEROV, Zh. I., ANDREEV, V. M., KOROL'KOV, V. I., PORTNOL, E. L., and TRET'YAKOV, D. N. (1968): *Fiz. Tekh. Poluprov*, **2**, pp. 1016–1017 [(1969): *Sov. Phys. – Semicond.*, **2**, pp. 843–844.]

ALFEROV, Zh. I., ANDREEV, V. M., GARBUZOV, D. Z., ZHILYAEV, Yu. V., MOROZOV, E. P., PORTNOI, E. L., and TROFIM, V. G. (1970): *Fiz. Tekh. Poluprov*, **4**, pp. 1826–1829 [(1971): *Sov. Phys. – Semicond.*, **4**, pp. 1573–1575.]

ANTYPAS, G. A. (1980): *Appl. Phys. Lett.*, **37**, pp. 64–65

ARAI, S., ASADA, M., TANBEN-EK, T., SUEMATSU, Y., ITAYA, Y., and KISHINO, K. (1981): *IEEE J. Quantum Electron.*, **QE-17**, pp. 640–645

ARNOLD, G., and PETERMANN, K. (1980): *Optical & Quantum Electron.*, **12**, pp. 207–219

ASADA, M., ADAMS, A. R., STUBKJAER, K. E., SUEMATSU, Y., ITAYA, Y., and ARAI, S. (1981): *IEEE J. Quantum Electron.*, **QE-17**, pp. 611–618

ASADA, M., and SUEMATSU, Y. (1982): *Appl. Phys. Lett.*, **41**, pp. 353–355

ASADA, M., and SUEMATSU, Y. (1983): *IEEE J. Quantum Electron.*, **QE-19**, pp. 917–923

ASAHI, H., KAWAMURA, Y., NOGUCHI, Y., MATSUOKA, T., and NAGAI, H. (1983): *Electron Lett.*, **19**, pp. 507–508

ASANO, T., OKUMURA, T., and AIZAWA, M. (1984): 9th IEEE Int. Semiconductor Laser Conference, Rio de Janeiro, pp. 42–3

BASOV, N. G., VUL, B., and POPOV, Yu. M. (1959): *Zh. Eksperim. Teor. Fiz.*, **37**, pp. 587–588 [(1960): *Sov. Phys. − JETP*, **10**, p. 416.]

BASOV, N. G., KROKHIN, O. N., and POPOV, Yu. M. (1961): *Zh. Eksperim Teor. Fiz.*, **40**, pp. 1879–1880 [(1961): *Sov. Phys. − JETP*, **13**, pp. 1320–1321.]

BEATTIE, A. R., and LANDSBERG, P. T. (1958): *Proc. Roy. Soc.*, **A249**, pp. 16–29

BERKEY, G. E. (1984): Tech. Digest of Conference on Optical Fiber Communication, 23–25 Jan. 1984, New Orleans, pp. 20–21

BERNARD, M. G. A., and DURAFFOURG, B. (1961): *Phys. Stat. Solidi*, **1**, pp. 699–703

BLONDEAU, R., RICCIARDI, J., HIRTZ, P., and DE CREMOUX, B. (1982): *Jap. J. Appl. Phys.*, **21**, pp. 1655

BLOW, K. J., and DORAN, N. J. (1983): *Electron. Lett.*, **19**, pp. 429–430

BLOW, K. J., DORAN, N. J., and CUMMINS, E. (1983): *Optics Communications*, **48**, pp. 181–184

BLUM, J. M., McGRODDY, J. C., McMULLIN, P. G., SHIH, K. K., SMITH, A. W., and ZIEGLER, J. F. (1975): *IEEE J. Quantum Electron.*, **QE-11**, pp. 413–418

BOERS, P. M., VLAARDINGERBROEK, M. T., and DANIELSEN, M. (1975): *Electron Lett.*, **11**, pp. 206–208

BOTEZ, D. (1978): *IEEE J. Quantum Electron.*, **QE-14**, pp. 230–232

BOTEZ, D. (1981): *IEEE J. Quantum Electron.*, **QE-17**, pp. 178–186

BOTEZ, D., and CONNOLLY, J. C. (1980): *Electron. Lett.*, **16**, pp. 942–4

BOTEZ, D., and HERSKOWITZ, G. J. (1980): *Proc. IEEE*, **68**, pp. 689–731

BOUADMA, N., RIOU, J., and BOULEY, J. C. (1982): *Electron. Lett.*, **18**, pp. 879–880

BOULEY, J. C., CHAMINANT, G., CHARIL, J., DEVOLDERE, P., and GILLERON, M. (1981): *Appl. Phys. Lett.*, **38**, pp. 845–847

BOWERS, J. E., and BURRUS, C. A. (1984): 9th IEEE Int. Semiconductor Laser Conference, Rio de Janeiro, Paper M1, pp. 158–159

BOWERS, J. E., HEMENWAY, B. R., BRIDGES, T. J., BURKHARDT, E. G., and WILT, D. P. (1985): *Electron. Lett.*, **21**, pp. 1090–1091

BROBERG, B., KOYAMA, F., TOHMORI, Y., and SUEMATSU, Y. (1984): *Electron. Lett.*, **20**, pp. 692–694

BROWN, G. N. (1986): *Br. Telecom Technol. J.*, **4**, pp. 71–80

BURKHARD, H. (1984): *J. Appl. Phys.*, **55**, pp. 503–508

BURKHARD, H., and KUPHAL, E. (1984): 9th Int. IEEE Semiconductor Laser Conference, Rio de Janeiro, Paper D-5, pp. 56–57

BURNHAM, R. D., STREIFER, W., SCIFRES, D. R., LINDSTROM, C., PAOLI, T. L., and HOLONYAK, N. (1982): *Electron Lett.*, **18**, pp. 1095–1097

BURRUS, C. A., LEE, T. P., DENTAI, A. G. (1981): *Electron. Lett.*, **17**, pp. 954–6

BURT, M. G. (1981): *J. Phys. C.*, **14**, 3269–3277

BURT, M. G. (1982): *Electron. Lett.*, **18**, pp. 806–807

BURT, M. G. (1983): *Electron. Lett.*, **19**, pp. 210–211

BUUS, J. (1981): *Appl. Optics*, **20**, pp. 1884–1885

BUUS, J. (1982): *IEEE J. Quantum Electron.*, **QE-18**, pp. 1083–1089

CAMERON, K. H., CHIDGEY, P. J., and PRESTON, K. R. (1982): *Electron. Lett.*, **18**, pp. 650–651

CAMPARO, J., and VOLK, C. H. (1982): *IEEE J. Quantum Electron*, **QE-18**, pp. 1990–1991

CASEY, H. C. Jr. (1984): *J. Appl. Phys.*, **56**, pp. 1959–1964

CASEY, H. C. Jr., and CARTER, P. L. (1984): *Appl. Phys. Lett.*, **44**, pp. 82–83
CASEY, H. C. Jr., and PANISH, M. B. (1978): 'Heterostructure Lasers — Pt. A: Fundamental Principles', New York (Academic, New York)
CASEY, H. C. Jr., and STERN, F. (1976): *J. Appl. Phys.*, **47**, pp. 631–643
CHANNIN, D. J., REDFIELD, D., and BOTEZ, D. (1984): 9th IEEE Int. Semiconductor Laser Conference, Rio de Janeiro, Paper I-1, pp. 112–113
CHATTERJEE, A. K., FAKTOR, M. M., LYONS, M. H., and MOSS, R. H. (1982): *J. Cryst. Growth*, **56**, pp. 591–604
CHEN, P. C., YU, K. L., MARGALIT, S., and YARIV, A. (1980): *Jap. J. Appl. Phys.*, **19**, pp. L775-L776
CHEN, P. C., YU, K. L., MARGALIT, S., and YARIV, A. (1981): *Appl. Phys. Lett.*, **38**, pp. 301–303
CHEN, T. R., KOREN, U., YU, K. L., LAU, K. Y., CHIU, L. C., HASSON, A., MARGALIT, S., and YARIV, A. (1982): *Appl. Phys. Lett.*, **41**, pp. 225–8
CHEN, T. R., CHIU, L. C., YU, K. L., KOREN, U., HASSON, A., MARGALIT, S., and YARIV, A. (1982a): *Appl. Phys. Lett.*, **41**, pp. 1115–1117
CHEN, T. R., CHAN, G. B., CHIU, L. C., YU, K. L., MARGALIT, S., and YARIV, A. (1983): *Appl. Phys. Lett.*, **43**, pp. 217–218
CHIN, A. K., ZIPFEL, C. L., GERA, M., CAMLIBEL, I., SKEATH, P., and CHIN, B. H. (1984): *Appl. Phys. Lett.*, **45**, pp. 37–39
CHINN, S. (1973): *IEEE J. Quantum Electron.*, **QE-9**, pp. 574–580
CHINONE, N., AIKI, K., NAKAMURA, M., and ITO, R. (1978): *IEEE J. Quantum Electron.*, **8**, pp. 625–631
CHIU, L. C., and YARIV, A. (1982): *IEEE J. Quantum Electron.*, **QE-18**, pp. 1406–1409
CHIU, L. C., CHEN, P. C., and YARIV, A. (1982): *IEEE J. Quantum Electron.*, **QE-18**, pp. 938–941
CHOI, H. K., and WANG, S. (1982): *Appl. Phys. Lett.*, **40**, pp. 571–3
CHOI, H. K., and WANG, S. (1983): *Electron. Lett.*, **19**, pp. 302–303
COHEN, L. G., MAMMEL, W. L., and JANG, S. J. (1982): *Electron. Lett.*, **18**, pp. 1023–1024
COLDREN, L. A., MILLER, B. I., IGA, K., and RENTSCHLER, J. A. (1981): *Appl. Phys. Lett.*, **5**, pp. 315–317
COTTER, D. (1983): *J. Opt. Commun.*, **4**, pp. 10–19
DAINO, B., SPANO, P., TAMBURRINI, M., and PIAZZOLLA, S. (1983): *IEEE J. Quantum Electron.*, **QE-19**, pp. 266–270
DANDRIDGE, A., and MILES, R. O. (1981): *Electron Lett.*, **17**, pp. 273–274
DANDRIDGE, A., and TAYLOR, H. F. (1982): *IEEE J. Quantum Electron.*, **QE-18**, pp. 1738–1750
DAVIES, G. J., HECKINGBOTTOM, R., OHNO, H., WOOD, C. E. C., and CALAWA, A. R. (1980): *Appl. Phys. Lett.*, **37**, pp. 290–292
DEVLIN, W. J., WALLING, R. H., FIDDYMENT, P. J., HOBBS, R. E., MURRELL, D., SPILLETT, R. E., and STEVENTON, A. G. (1981): *Electron Lett.*, **17**, pp. 651–653
DIDCHENKO, R., ALIX, J. E., and TOENISKOTLER, R. H. (1960): *J. Inorg. & Nucl. Chem.*, **14**, pp. 35–37
DINGLE, R. (1975): Festkorperprobleme XV', QUEISSER, H. J. (Ed.) (Pergamon, New York) pp. 21–48
DIXON, R. W., and JOYCE, W. B. (1979) *J. Appl. Phys.*, **50**, pp. 4591–4595
DOI, A., CHINONE, N., AIKI, K., and ITO, R. (1979): *Appl. Phys. Lett.*, **34**, pp. 393–395
DOI, A., FUKUZAWA, T., NAKAMURA, M., ITO, R., and AIKI, K. (1979a): *Appl. Phys. Lett.*, **35**, pp. 441–3
DOLGINOV, L. M., DRAKIN, A. E., DRUZHININA, L. V., ELISEEV, P. G., MILVIDSKY, M. G., SKRIPKIN, V. A., and SVERDLOV, B. N. (1981): *IEEE J. Quantum Electron.*, **QE-17**, pp. 593–596
DOUSMANIS, G. C., NELSON, H., and STAEBLER, D. L. (1964): *Appl. Phys. Lett.*, **5**,

pp. 174–176

DUMKE, W. P. (1962): *Phys. Rev.*, **127**, pp. 1559–1565

DUTTA, N. K. (1980): *J. Appl. Phys.*, **51**, pp. 6095–6100

DUTTA, N. K. (1981): *J. Appl. Phys.*, **52**, pp. 55–60

DUTTA, N. K. (1983): *J. Appl. Phys.*, **54**, pp. 1236–1245

DUTTA, N. K., and NELSON, R. J. (1982): *J. Appl. Phys.*, **53**, pp. 74–92

DUTTA, N. K., and NELSON, R. J. (1982a): *IEEE J. Quantum Electron.*, **QE-18**, 44–49

DUTTA, N. K., HARTMAN, R. L., and TSANG, W. T. (1983): *IEEE J. Quantum Electron.*, **QE-19**, pp. 1243–1246

DUTTA, N. K., NAPHOLTZ, S. G., WILSON, R. B., BROWN, R. L., CELLA, T., CRAFT, D. C. (1984): *Appl. Phys. Lett.*, **45**, pp. 941–943

DYMENT, J. C., and D'ASARO, L. A. (1967): *Appl. Phys. Lett.*, **11**, pp. 292–294

DYMENT, J. C., D'ASARO, L. A., NORTH, J. C., MILLER, B. I., and RIPPER, J. E. (1972): *Proc. IEEE*, **60**, pp. 726–728

DYOTT, R. B. (1986): *IEE Proc.* Pt. J, **133**, pp. 199–201

EAGLESFIELD, C. C. (1962): *Proc. IEE*, **109B**, pp. 26–32

EBELING, K. J., COLDREN, L. A., MILLER, B. I., and RENTSCHLER, J. A. (1983): *Appl. Phys. Lett.*, **42**, pp. 6–8

EISENSTEIN, G., KOREN, U., TUCKER, R. S., KASPER, B. L., GNAUCK, A. H., and TIEN, P. K. (1984): *Appl. Phys. Lett.*, **45**, pp. 311–313

ELSÄSSER, W., GOBEL, E. O., and KUHL, J. (1983): *IEEE J. Quantum Electron.*, **QE-19**, pp. 981–985

ELSÄSSER, W., and GOBEL, E. O. (1984a): *Appl. Phys. Lett.*, **45**, pp. 353–355

ENDO, K., MATSUMOTO, S., KAWANO, H., SAKUMA, I., and KAMEJIMA, T. (1982): *Appl. Phys. Lett.*, **40**, pp. 921–923

EPWORTH, R. E. (1978): 4th European Conference on Optical Communications, Genoa, pp. 492–501

FATTAH, I. H. A., and WANG, S. (1982): *Appl. Phys. Lett.*, **41**, pp. 112–4

FIGUEROA, A. L., SLAYMAN, C. W., and YEN, H. W. (1982): *IEEE J. Quantum Electron.*, **QE-18**, pp. 1718–1727

FLEMING, M. W., and MOORADIAN, A. (1981): *Appl. Phys. Lett.*, **38**, pp. 511–513

FORREST, S. R. (1985): *IEEE J. Lightwave Technol.*, **LT-3**, pp. 1248–1263

FRISCH, D. A., and HENNING, I. D. (1982): *Electron. Lett.*, **18**, pp. 129–130

FRISCH, D. A., and HENNING, I. D. (1984): *Electron. Lett.*, **20**, pp. 631–632

FUJITA, T., OHYA, J., MATSUDA, K., ISHINO, M., SATO, H., and SERIZAWA, H. (1985): *Electron Lett.*, **21**, pp. 374–376

FUKUDA, M. (1983): *IEEE J. Quantum. Electron*, **QE-19**, pp. 1692–1698

FUKUDA, M., TAKAHEI, K., IWANE, G., and IKEGAMI, T. (1982): *Appl. Phys. Lett.*, **41**, pp. 18–21

FUKUDA, M., FUJITA, O., and IWANE, G. (1984): *IEEE Trans.*, **CHMT-7**, pp. 202–206

FURUYA, K., SUEMATSU, Y., and HONG, T. (1978): *Appl. Optics*, **17**, pp. 1949–1952

GAMBLING, W. A. (1986): *IEE Proc.* Pt. J, **133**, pp. 205–210

GARMIRE, E., EVANS, G., NILSEN, J. (1981): *Appl. Phys. Lett.*, **39**, pp. 789–791

GARRETT, I. (1983): *IEEE J. Lightwave Technol.*, **LT-1**, pp. 131–138

GILLING, L. J. (1982): *J. Electrochem. Soc.*, **129**, pp. 634–637

GÖBEL, E. O. (1982): *in* 'GaInAsP alloy semiconductors', Pearsall, T. P. (Ed.) (Wiley) pp. 313–338

GOEBEL, E. O., LUZ, G., and SCHLOSSER, E. (1979): *IEEE J. Quantum Electron.*, **QE-15**, pp. 697–700

GOODMAN, C. H. L. (1978): *IEEE J. Solid State & Electron Devices*, **2**, pp. 129–137

GOUBAU, G., and SCHWERING, F. (1961): *IRE Trans.*, **AP-9**, pp. 248–256

GROOVES, S. H., and PLONKO, M. C. (1981): *Appl. Phys. Lett.*, **38**, pp. 1003–1004

HAKKI, B. W., and PAOLI, T. L. (1975): *J. Appl. Phys.*, **46**, pp. 1299–1306

HALL, R. N. (1976): *IEEE Trans.*, **ED-23**, pp. 700–704

HALL, R. N., FENNER, G. E., KINGSLEY, J. D., SOLTYS, T. J., and CARLSON, R. O. (1962): *Phys. Rev. Lett.*, **9**, pp. 366–368

HALLIWELL, M. A. G., LYONS, M. H., and HILL, M. J. (1984): *J. Cryst. Growth*, **68**, pp. 523–531

HARDER, C., VAHALA, K., and YARIV, A. (1983): *Appl. Phys. Lett.*, **42**, pp. 328–330

HARMON, R. A. (1982): *Electron. Lett.*, **18**, pp. 1058–1060

HASEGAWA, A., and TAPPERT, F. (1973): *Appl. Phys. Lett.*, **23**, pp. 142–144

HATCH, C. B., MURRELL, D. L., and WALLING, R. H. (1982): *IEE Proc.* Pt. I, **129**, pp. 214–217

HAUG, A. (1983): *Appl. Phys. Lett.*, **42**, pp. 512–514

HAUS, H. A., and SHANK, C. V. (1976): *IEEE J. Quantum Electron.*, **QE-12**, pp. 532–539

HAYASHI, I., PANISH, M. B., and FOY, P. W. (1969): *IEEE J. Quantum Electron.*, **QE-5**, pp. 211–212

HAYASHI, I., PANISH, M. B., FOY, P. W., and SUMSKI, S. (1970): *Appl. Phys. Lett.*, **17**, pp. 109–111

HENNING, I. D. (1982): *Electron. Lett.*, **18**, pp. 368–369

HENNING, I. D. (1983): *Electron Lett.*, **19**, pp. 935–937

HENNING, I. D. (1984): *IEEE Proc.* Pt. H, **131**, pp. 133–138

HENNING, I. D., and COLLINS, J. V. (1983): *Electron. Lett.*, **19**, pp. 927–929

HENNING, I. D., and FRISCH, D. A. (1983): *IEEE J. Lightwave Technol.*, **LT-1**, pp. 202–206

HENNING, I. D., WESTBROOK, L. D., NELSON, A. W., and FIDDYMENT, P. J. (1984): *Electron. Lett.*, **20**, pp. 885–887

HENRY, C. H. (1982): *IEEE J. Quantum Electron.*, **QE-18**, pp. 259–264

HENRY, C. H. (1983): *IEEE J. Quantum Electron.*, **QE-19**, pp. 1391–1397

HENRY, C. H., LOGAN, R. A., and MERRITT, F. R. (1980): *J. Appl. Phys.*, **51**, pp. 3042–3050

HENRY, C. H., LOGAN, R. A., TEMKIN, H., and MERRITT, F. R. (1983): *IEEE J. Quantum Electron.*, **QE-19**, pp. 941–946

HENRY, C. H., LOGAN, R. A., MERRITT, F. R., and LUONGO, J. P. (1983a): *IEEE J. Quantum Electron.*, **QE-19**, pp. 947–952

HERSEE, S., BALDY, M., ASSENAT, P., DE CREMOUX, B., and DUCHEMIN, J. P. (1982): *Electron. Lett.*, **18**, pp. 618–620

HESS, K., VOJAK, B. A., HOLONYAK, N., Jr., CHIN, R., and DAPKUS, P. D. (1980): *Solid State Electron.*, **23**, pp. 585–589

HIGUCHI, H., OOMURA, E., HIRANO, R., SAKAKIBARA, Y., NAMIZAKI, H., SUSAKI, W., and FUJIKAWA, K. (1983): *Electron Lett.*, **19**, pp. 976–977

HILSUM, C. (1969): *in* 'Gallium arsenide lasers', Gooch, C. H. (Ed.) (Wiley, Chichester)

HIRANO, R., OOMURA, E., HIGUCHI, H., SAKAKIBARA, Y., NAMIZAKI, H., SUSAKI, W. (1982): Digest of Technical papers of the 14th Conference on Solid State Devices, Tokyo, pp. 27–28

HIRANO, R., OOMURA, E., HIGUCHI, E., SAKAKIBARA, Y., NAMIZAKI, H., SUSAKI, W., and FUJIKAWA, K. (1983): *Appl. Phys. Lett.*, **43**, pp. 187–189

HIRAO, M., DOI, A., TSUJI, S., NAKAMURA, M., and AIKI, K. (1980): *J. Appl. Phys.*, **51**, pp. 4539–4540

HODGKINSON, T. G., WYATT, R., and SMITH, D. R. (1982): *Electron. Lett.*, **18**, pp. 523–525

HOLONYAK, N., Jr, and BEVACQUA, S. F. (1962): *Appl. Phys. Lett.*, **1**, pp. 82–82

HOLONYAK, N., Jr, KOLBAS, R. M., DUPUIS, R. D., and DAPKUS, P. D. (1980): *IEEE J. Quantum Electron.*, **QE-16**, pp. 170–185

'tHOOFT, G. W. (1981): *Appl. Phys. Lett.*, **39**, pp. 389–390

HOOPER, R. C., MIDWINTER, J. E., SMITH, D. W., and STANLEY, I. W. (1983): *IEEE J. Lightwave Technol.*, **LT-1**, pp. 596-611

HORIKAWA, H., IMANAKA, K., MATOBA, A., KAWAI, Y., and SAKUTA, M. (1984): *Appl. Phys. Lett.*, **45**, pp. 328-330

HORIKOSHI, Y., and FURUKAWA, Y. (1979): *Japan J. Appl. Phys.*, **18**, pp. 809-815

HSIEH, J. J. (1979): *IEEE J. Quantum Electron.*, **QE-15**, pp. 694-697

HSIEH, J. J., ROSSI, J. A., and DONNELLY, J. P. (1976): *Appl. Phys. Lett.*, **28**, pp. 709-711

HSIEH, J. J., and SHEN, C. C. (1977): *Appl. Phys. Lett.*, **30**, pp. 429-431

IGA, K., SODA, H., TERAKADO, T., and SHIMIZU, S. (1983): *Electron. Lett.*, **19**, pp. 457-458

IKEGAMI, T., TAKAHEI, K., FUKUDA, M., and KUROIWA, K. (1983): *Electron. Lett.*, **19**, pp. 282-283

IKUSHIMA, I., and MAEDA, M. (1979): *IEEE J. Quantum Electron.*, **QE-15**, pp. 844-845

IMANAKA, K., HORIKAWA, H., MATOBA, A., KAWAI, Y., and SAKUTA, M. (1984): *Appl. Phys. Lett.*, **45**, pp. 282-283

ISHIKAWA, H., IMAI, H., TANAHASHI, T., NISHITONI, Y., and TAKUSAGAWA, M. (1981): *Electron. Lett.*, **17**, pp. 465-467

ITAYA, Y., KATAYAMA, S., and SUEMATSU, Y. (1979): *Electron. Lett.*, **15**, pp. 123-124

ITAYA, Y., MATSUOKA, T., NAKAND, Y., SUZUKI, Y., KUROIWA, K., and IKEGAMI, T. (1982): *Electron. Lett.*, **18**, pp. 1006-1008

ITAYA, Y., MATSUOKA, T., KUROIWA, K., and IKEGAMI, T. (1983): 4th Int. Conference on Integrated Optics and Optical Fiber Communications, Tokyo, pp. 154-155

ITAYA, Y., MATSUOKA, T., KUROIWA, K., and IKEGAMI, T. (1984): *IEEE J. Quantum Electron.*, **QE-20**, pp. 230-235

ITO, H., ONODERA, N., GEN-EI, K., and INABA, H. (1981): *Electron. Lett.*, **17**, pp. 15-16

ITO, T., MACHIDA, S., NAWATA, K., and IKEGAMI, T. (1977): *IEEE J. Quantum Electron.*, **QE-13**, pp. 574-579

JACKEL, H., and GUEKOS, G. (1977): *Opt. & Quantum Electron.*, **9**, pp. 233-239

JOHNSON, L. F., KAMMLOTT, G. W., and INGERSOLL, K. A. (1968): *Appl. Opt.*, **17**, pp. 1165-1181

KAMINOW, I. P. (1984): Conference on Optical Fiber Communication, New Orleans, 1984, Paper MJ1

KAMINOW, I. P., NAHORY, R. E., POLLACK, M. A., STULZ, L. W., and DeWINTER, J. C. (1979): *Electron. Lett.*, **15**, pp. 763-764

KANO, H., OE, K., ANDO, S., and SUGIYAMA, K. (1978): *Japan J. Appl. Phys.*, **17**, pp. 1887-1888

KANO, H., and SUGIYAMA, K. (1979): *J. Appl. Phys.*, **50**, pp. 7934-7938

KAO, K. C., and HOCKHAM, G. A. (1966): *Proc. IEE*, **113**, 1151-1158

KAPRON, F. P. (1977): *Electron. Lett.*, **13**, pp. 96-97

KAPRON, F. P., KECK, D. B., and MAURER, R. D. (1970): *Appl. Phys. Lett.*, **17**, pp. 423-425

KARBOWIAK, A. E. (1986): *IEE Proc.*, Pt. J, **133**, pp. 202-204

KASAHARA, K., TERAKADO, T., SUZUKI, A., and MURATA, S. (1986): *IEEE J. Lightwave Technol.*, **LT-4**, pp. 907-912

KASEMSET, D., HONG, C. S., PATEL, N. B., and DAPKUS, P. D. (1982): *Appl. Phys. Lett.*, **41**, pp. 912-914

KASEMSET, D., HONG, C. S., PATEL, N. B., and DAPKUS, P. D. (1983): *IEEE J. Quantum Electron.*, **QE-19**, pp. 1025-1029

KATAOKA, H., and IKEDA, M. (1984): *Electron Lett.*, **20**, pp. 438-439

KAWACHI, M., KAWANA, A., and MIYASHITA, T. (1977): *Electron. Lett.*, **13**, pp. 442-443

KAWAMURA, K., and YAMAMOTO, T. (1976): *J. Cryst. Growth*, **32**, pp. 157-163

KAWAMURA, Y., NOGUCHI, Y., ASAHI, H., and NAGAI, H. (1982): *Electron. Lett.*, **18**, pp. 91-92

KAWAMURA, Y., WAKITA, K., ITAYA, Y., YOSHIKUNI, Y., and ASAHI, H. (1986): *Electron. Lett.*, **22**, pp. 242–243

KAWANISHI, H., SUEMATSU, Y., UTAKA, K., ITAYA, Y., and ARAI, S. (1979): *IEEE J. Quantum Electron.*, **QE-15**, pp. 701–706

KHOE, G. D., POULISSEN, J., and DeVRIEZE, H. M. (1983): *Electron Lett.*, **19**, pp. 205–207

KIKUCHI, K., OKOSHI, T., and KAWAI, T. (1984): *Electron. Lett.*, **20**, pp. 450–451

KIKUCHI, K., OKOSHI, T., and ARATA, R. (1984a): *Electron Lett.*, **20**, pp. 535–536

KIM, S.-H., and FONSTAD, C. G. (1979): *IEEE J. Quantum Electron.*, **QE-15**, pp. 1405–1408

KINOSHITA, J., OKUDA, H., and UEMATSU, Y. (1983): *Electron. Lett.*, **19**, pp. 215–216

KISHINO, K., SUEMATSU, Y., and ITAYA, Y. (1979): *Electron. Lett.*, **15**, pp. 134–6

KITAMURA, M., SEKI, M., YAMAGUCHI, M., ITO, I., KOBAYASHI, Ke, KOBAYASHI, Ko., and MATSUOKA, T. (1983): *Electron Lett.*, **19**, pp. 840–841

KITAMURA, M., YAMAGUCHI, M., MURATA, S., MITO, I., and KOBAYASHI, K. (1984): *Electron. Lett.*, **20**, pp. 595–596

KOBAYASHI, N., and HORIKOSHI, Y. (1979): *Jap. J. Appl. Phys.*, **18**, pp. 1005–1006

KOBAYASHI, N., HORIKOSHI, Y., and UEMURA, C. (1980): *Jap. J. Appl. Phys.*, **19**, Suppl. 19-3, pp. 333–339

KOBAYASHI, K., UTAKA, K., ABE, Y., and SUEMATSU, Y. (1981): *Electron. Lett.*, **17**, pp. 366–368

KOBAYASHI, H., IWAMURA, H., SAKU, T., and OTSUKA, K. (1983): *Electron. Lett.*, **19**, pp. 166–168

KOCH, T. L., COLDREN, L. A., BRIDGES, T. J., BURKHARDT, E. G., CORVINI, P. J., WILT, D. P., and MILLER, B. I. (1984): 9th IEEE Int. Semiconductor Laser Conference, Rio de Janeiro, pp. 80–81

KODAMA, K., OZEKI, M., and KOMENO, J. (1984): *Electron. Lett.*, **20**, pp. 48–50

KOGELNIK, H., and SHANK, C. V. (1972): *J. Appl. Phys.*, **43**, pp. 2327–2335

KOJIMA, K., and KYUMA, K. (1984): *Electron. Lett.*, **20**, pp. 869–871

KOMENO, J., TAKIKAWA, M., and OZEKI, M. (1983): *Electron. Lett.*, **19**, pp. 473–474

KOMIYA, S., YAMAZAKI, S., KISHI, Y., UMEBU, I., and KOTANI, T. (1983): *J. Cryst. Growth*, **61**, pp. 232–239

KONNERTH, K., and LANZA, C. (1964): *Appl. Phys. Lett.*, **4**, pp. 120–121

KOREN, U., CHEN, T. R., HARDER, C., HASSON, A., YU, K. L., CHIU, L. C., MARGALIT, S., and YARIV, A. (1983): *Appl. Phys. Lett.*, **42**, pp. 403–405

KOREN, U., RAV-NOY, Z., HASSON, A., CHEN, T. R., YU, K. L., CHIU, L. C., MARGALIT, S., and YARIV, A. (1983a): *Appl. Phys. Lett.*, **42**, pp. 848–850

KOREN, U., ARAI, S., and TIEN, P. K. (1984): *Electron. Lett.*, **20**, pp. 177–8

KOYAMA, F., SUEMATSU, Y., KOJIMA, K., and FURUYA, K. (1984): *Electron. Lett.*, **20**, pp. 391–392

KRESSEL, H., and NELSON, H. (1969): *RCA Rev.*, **30**, pp. 106–113

KRESSEL, H., NELSON, H., and HAWRYLO, F. Z. (1969): *J. Appl. Phys.*, **41**, pp. 2019–2031

KROEMER, H. (1963): *Proc. IEEE*, **51**, pp. 1782–1783

LANDSBERG, P. T., ABRAHAMS, M. S., and OSINSKI, M. (1984): *IEEE J. Quantum Electron.*, **QE-21**, pp. 24–28

LASHER, G. J., and STERN, F. (1964): *Phys. Rev.*, **A133**, pp. 553–563

LAU, K. Y., BAR-CHAIM, N., HARDER, Ch., and YARIV, A. (1983a): *Appl. Phys. Lett.*, **43**, pp. 329–331

LAU, K. Y., BAR-CHAIM, N., URY, N., and YARIV, A. (1984): Conference on Optical Fiber Communication, New Orleans, Postdeadline paper WJ1

LEBURTON, J. P., and HESS, K. (1983): *J. Vac. Sci. Technol.*, **B1**, pp. 415–419

LEE, T. P. (1975): *Bell Syst. Tech. J.*, **54**, pp. 53–68

LEE, T. P., BURRUS, C. A., MILLER, B. I., and LOGAN, R. A. (1975): *IEEE J. Quantum Electron.*, **QE-11**, pp. 432–435

LEE, T. P., BURRUS, C. A., COPELAND, J. A., DENTAI, A. G., and MARCUSE, D. (1982): *IEEE J. Quantum Electron*, **QE-18**, pp. 1101–1112

LEE, T. P., BURRUS, C. A., LIU, P. L., and DENTAI, A. G. (1982a): *Electron Lett.*, **18**, pp. 805–806

LEE, T. P., BURRUS, C. A., LINKE, R. A., and NELSON, R. J. (1983): *Electron. Lett.*, **19**, pp. 82–84

LEE, T. P., BURRUS, C. A., LIOU, K. Y., OLSSON, N. A., LOGAN, R. A., and WILT, D. P. (1984): *Electron. Lett.*, **20**, 1011–1012

LEHENY, R. F., NAHORY, R. E., POLLACK, M. A., BALLMAN, A. A., BEEBE, E. D., DeWINTER, J. C., and MARTIN, R. J. (1980): *Electron Lett.*, **16**, pp. 353–355

LIAU, Z. L., and WALPOLE, J. N. (1982): *Appl. Phys. Lett.*, **40**, pp. 568–570

LIAU, Z. L., WALPOLE, J. N., FLANDERS, D. C., TSANG, D. Z., and DeMEO, N. L. (1984): 9th IEEE Int. Semiconductor Laser Conference, Brazil, Post deadline Paper N5

LIN, B. J. (1983): in 'Introduction to Microlithography' Thompson, L. F., Wilson, C. G., and Bowden, M. J. (Eds.) Amer. Chem. Society Series 219, Washington, DC, chap. 6

LINKE, R. A. (1984): *Electron. Lett.*, **20**, pp. 472–474

LINKE, R. A., KASPER, B. L., CAMPBELL, J. C., DENTAI, A. G., and KAMINOW, I. P. (1984): Conference on Optical Fiber Communication, 23–25 January, New Orleans, Post deadline paper WJ7

LINKE, R. A., KASPER, B. L., JOPSON, R. M., CAMPBELL, J. C., DENTAI, A. G., TSANG, W. T., OLSSON, N. A., LOGAN, R. A., JOHNSON, L. S., and HENRY, C. H. (1984a): 9th Int. IEEE Conference on Semiconductor Lasers, Rio de Janeiro, pp. 70–71

LIOU, K. Y., BURRUS, C. A., LINKE, R. A., KAMINOW, I. P., GRANLUND, S. W., SWAN, C. B., and BESOMI, P. (1984): *Appl. Phys. Lett.*, **45**, pp. 729–731

LOCKWOOD, H. F., KRESSEL, H., SOMMERS, H. S. Jr., and HAWRYLO, F. Z. (1970): *Appl. Phys. Lett.*, **17**, pp. 499–502

MALYON, D. J., and McDONNA, A. P. (1982): *Electron. Lett.*, **18**, pp. 445–447

MANNING, J., OLSHANSKY, R., and SU, C. B. (1983): *IEEE J. Quantum Electron.*, **QE-19**, pp. 1525–1530

MARSHALL, I. W., O'MAHONY, M. J., and CONSTANTINE, P. D. (1986): *Electron. Lett.*, **22**, pp. 253–255

MASUEDA, H., SASAI, S., and NAKAMURA, M. (1983): *IEEE J. Lightwave Technol.*, **LT-1**, pp. 261–269

MATSUNAGA, T., and IKEDA, M. (1985): *Electron. Lett.*, **21**, pp. 945–946

MATSUOKA, T., NAGAI, H., NOGUCHI, Y., SUZUKI, Y., and KAWAGUCHI, Y. (1984): *Japan J. Appl. Phys.*, **23**, L138-L140

McCUMBER, D. E. (1966): *Phys. Rev.*, **141**, pp. 306–322

MELCHIOR, H. (1980): Proc. Top. Meeting on Integrated & Guided-Wave Optics, Paper MA2

MERTON, R. K. (1961): *New Scientist*, **259**, pp. 306–308

MIKAMI, O. (1981): *Japan J. Appl. Phys.*, **20**, L488-L490

MIKAMI, O., NAKAGOME, H., YAMAUCHI, Y., and KANBE, H. (1982): *Electron. Lett.*, **18**, pp. 237–239

MILLER, R. C., KLEINMAN, D. A., and GOSSARD, A. C. (1984): *Phys. Rev.*, **B 29**, pp. 7085–7087

MITO, I., KATAMURA, M., KAEDE, K., ODAGIRI, Y., SEKI, M., SUGIMOTO, M., and KOBAYASHI, K. (1982): *Electron Lett.*, **18**, pp. 2–3

MITO, I., KATAMURA, M., KOBAYASHI, Ke, and KOBAYASHI, Ko (1982a): *Electron Lett.*, **18**, pp. 953–954

MITO, I., KITAMURA, M., KOBAYASHI, K., MURATA, S., SEKI, M., ODAGIRI, Y., NISHIMOTO, H., YAMAGUCHI, M., and KOBAYASHI, K. (1983): *IEEE J. Lightwave Technol.*, **LT-1**, pp. 195–201

MIYA, T., TERUNUMA, Y., HOSAKA, T., and MIYASHITA, T. (1979): *Electron. Lett.*, **15**, pp. 106–108

MIZUISHI, K., HIRAO, M., TSUJI, S., SATO, H., and NAKAMURA, M. (1980): *Jap. J. Appl. Phys.*, **19**, pp. L429–432

MIZUISHI, K.-I., SAWAI, M., TODOROKI, S., TSUJI, S., HIRAO, M., and NAKAMURA, M. (1983): *IEEE J. Quantum Electron.*, **QE-19**, pp. 1294–1301

MIZUTANI, T., YOSHIDA, M., USAI, A., WATANABE, H., YUASA, T., and HAYASHI, I. (1980): *Jap. J. Appl. Phys.*, **19**, pp. L113-L116

MOLLENAUER, L. F., STOLEN, R. H., and GORDON, J. P. (1980): *Phys. Rev. Lett.*, **45**, pp. 1095–1098

MORIKI, K., WAKAO, K., KITAMURA, M., IGA, K., and SUEMATSU, Y. (1980): *Jap. J. Appl. Phys.*, **19**, pp. 2191–2196

MOSS, R. H. (1984): *J. Cryst. Growth*, **68**, pp. 78–87

MOSS, R. H., and EVANS, J. S. (1981): *J. Cryst. Growth*., **55**, pp. 129–134

MOZER, A., ROMANEK, K. M., SCHMID, W., PILKUHN, M. H., and SCHLOSSER, E. (1982): *Appl. Phys. Lett.*, **41**, pp. 964–966

MOZER, A., MENNER, R., and PILKUHN, M. H. (1983): 4th International Conference on Integrated Optics and Optical Fiber Communication, Tokyo, pp. 218–219

MUKAI, T., YAMAMOTO, Y., and KIMURA, T. (1983): *Rev. Elect. Commun. Lab.*, **31**, pp. 340–348

MUKAI, T., and YAMAMOTO, Y. (1984): *Electron. Lett.*, **20**, pp. 29–30

MURATA, S., KITAMURA, M., YAMAGUCHI, M., MITO, I., and KOBAYASHI, K. (1983): *Electron. Lett.*, **19**, pp. 1084–1085

MUROTANI, T., OOMERA, E., HIGUCHI, H., NAMIZAKI, H., and SUSAKI, W. (1980): *Electron. Lett.*, **16**, pp. 566–568

MURRELL, D. L., WALLING, R. H., HOBBS, R. E., and DEVLIN, W. J. (1983): *IEE Proc.* Pt. I, **129**, pp. 209–213

NAGAI, H., and NOGUCHI, Y. (1978): *Appl. Phys. Lett.*, **32**, pp. 234–236

NAGAI, H., NOGUCHI, Y., TAKEHEI, K., TOYOSHIMA, Y., and IWANE, G. (1980): *Jap. J. Appl. Phys.*, **19**, pp. L218-L220

NAGAI, H., NOGUCHI, Y., MATSUOKA, T., and SUZUKI, Y. (1983): *Jap. J. Appl. Phys.*, **22**, pp. L291-L293

NAHORY, R. E., and POLLACK, M. A. (1978): *Electron. Lett.*, **14**, pp. 727–729

NAHORY, R. E., POLLACK, M. A., BEEBE, E. D., DeWINTER, J. C., and DIXON, R. W. (1976): *Appl. Phys. Lett.*, **28**, pp. 19–21

NAKAMURA, H., AIKI, R., CHINONE, R. I., and UNEO, A. J. (1978): *J. Appl. Phys.*, **49**, pp. 4644–4648

NAKANO, Y., TAKAHEI, K., NOGUCHI, Y., SUZUKI, Y., NAGAI, H., and NAWATA, K. (1981): *Electron. Lett.*, **17**, pp. 782–783

NAKANO, Y., TAKAHEI, K., NOGUCHI, Y., SUZUKI, Y., and NAGAI, H. (1982): *Electron. Lett.*, **18**, pp. 501–502

NAKANO, Y., IWANE, G., and IKEGAMI, T. (1984): *Electron. Lett.*, **20**, pp. 397–398

NAMIZAKI, H., KAN, H., ISHII, M., and ITO, A. (1974): *J. Appl. Phys.*, **45**, pp. 2785–2786

NASH, F. R., and HARTMAN, R. L. (1979): *J. Appl. Phys.*, **50**, pp. 3133–3141

NATHAN, M. I., DUMKE, W. P., BURNS, G., DILL, F. H., Jr, and LASHER, G. (1962): *Appl. Phys. Lett.*, **1**, pp. 62–64

NELSON, A. W., and WHITE, E. A. D. (1982): *J. Cryst. Growth*, **57**, pp. 610–612

NELSON, A. W., WESTBROOK, L. D., and EVANS, J. S. (1983): *Electron. Lett.*, **19**, pp. 34–36

NELSON, A. W., and WESTBROOK, L. D. (1984): *J. Cryst. Growth*, **68**, pp. 102–110

NELSON, B. P., COTTER, D., BLOW, K. J., and DORAN, N. J. (1983): *Optics Communications*, **48**, pp. 292–294

NELSON, H. (1963): *RCA Rev.*, **24**, pp. 603–626

NELSON, R. J., WRIGHT, P. D., BARNES, P. A., BROWN, R. L., CELLA, T., and SOBERS, R. G. (1980): *Appl. Phys. Lett*, **36**, pp. 358–360

NISHI, H., YANO, M., NISHITANI, Y., ALISTA, Y., and TAKUSAGAWA, M. (1979): *Appl. Phys. Lett.*, **35**, pp. 232–234

NOMURA *et al.* (1981): Ref. 40 in ARAI, S., ITAYA, Y., KISHINO, K., SUEMATSU, Y., MORIKI, K., WAKAO, K., and IGA, K. (1982): 'Long-wavelength semiconductor lasers' *in* Japan Annual Reviews in Electronics Computers and Telecommunications. Suematsu, Y. (Ed.) Optical devices and fibers (North Holland, 1982)

NUESE, C. J., and OLSEN, G. H. (1975): *Appl. Phys. Lett.*, **26**, pp. 528–531

OE, K., ANDO, S., and SUGIYAMA, K. (1980): *J. Appl. Phys.*, **51**, pp. 43–49

OGAWA, K. (1982): *IEEE J. Quantum Electron.*, **QE-18**, pp. 849–855

OKANO, Y., KAKAGAWA, K., and ITO, T. (1980): *IEEE Trans.*, **COM-28**, pp. 238–243

OKUDA, H., KINOSHITA, J., HIRAYAMA, Y., and UEMATSU, Y. (1983): *Electron. Lett.*, **19**, pp. 362–363

OKUDA, H., HIRAYAMA, Y., FURUYAMA, H., UEMATSU, Y., and BEPPU, T. (1984): 9th IEEE Int. Semiconductor Laser Conference, Rio de Janeiro, pp. 178–179

OLESEN, H., SAITO, S., MUKAI, T., SAITOH, T., and MIKAMI, O. (1983): *Jap. J. Appl. Phys.*, **22**, L664–666

OLSHANSKY, R., SU, C. B., MANNING, J., and SCHLAFER, J. (1983): *Electron. Lett.*, **19**, pp. 867–868

OLSHANSKY, R., SU, C. B., MANNING, J., and POWAZINIK, W. (1984): *IEEE J. Quantum Electron.*, **QE-20**, pp. 838–854

OLSHANSKY, R., LANZISERA, V., SU, C. B., POWAZINIK, W., and LAUER, R. B. (1986): *Appl. Phys. Lett.*, **49**, pp. 128–130

OLSSON, N. A., DUTTA, N. K., and LIOU, K. Y. (1984): *Electron. Lett.*, **20**, pp. 121–122

O'MAHONY, M. J., MARSHALL, I. W., WESTLAKE, H. J., and STALLARD, W. G. (1986): *Electron. Lett.*, **22**, p. 1238

O'MAHONY, M. J. (1987): *IEEE J. Lightwave Technol.*, submitted

OOMURA, E., MUROTANI, T., HIGUCHI, H., NAMIZAKI, H., and SUSAKI, W. (1981): *IEEE J. Quantum Electron.*, **QE-17**, pp. 646–650

OOMURA, E., HIGUCHI, H., HIRANO, R., SAKAKIBARA, Y., NAMIZAKI, H., and SUSAKI, W. (1983): *Electron. Lett.*, **19**, pp. 407–408

OSINSKI, M., and ADAMS, M. J. (1982): *IEE Proc.* Pt. I, **129**, pp. 229–236

OTSUKA, K., and TARUCHA, S. (1981): *IEEE J. Quantum Electron.*, **QE-17**, pp. 1515–1521

PANISH, M. B. (1980): *J. Electrochem. Soc.*, **127**, pp. 2729–2733

PANISH, M. B., CASEY, H. C., Jr, SUMSKI, S., and FOY, P. W. (1973): *Appl. Phys. Lett.*, **22**, pp. 590–591

PANISH, M. B., and TEMKIN, H. (1984): *Appl. Phys. Lett.*, **44**, pp. 785–787

PAOLI, T., HAKKI, B. W., and MILLER, B. I. (1973): *J. Appl. Phys.*, **44**, pp. 1276–1280

PARTIN, D. L. (1984): *Appl. Phys. Lett.*, **45**, pp. 487–489

PAYNE, D. N., and GAMBLING, W. A. (1975): *Electron. Lett.*, **11**, pp. 176–178

PEOPLE, R., WECHT, K. W., ALAVI, K., and CHO, A. Y. (1983): *Appl. Phys. Lett.*, **43**, pp. 118–120

PETERMAN, K. (1978): *Opt. & Quantum Electron.*, **10**, pp. 233–242

PETTIT, G. D., and TURNER, W. J. (1965): *J. Appl. Phys.*, **36**, p. 2081

PLASTOW, R., HARDING, M., GRIFFITH, I., CARTER, A. C., and GOODFELLOW, R. C. (1982): *Electron. Lett.*, **18**, pp. 262–263

PRESTON, K. R., WOLLARD, K. C., and CAMERON, K. H. (1981): *Electron. Lett.*, **17**, pp. 931–932

PRINCE, F. C., PATEL, N. B., and BULL, D. J. (1980): *IEEE J. Quantum Electron.*, **QE-16**, pp. 1034–1038

PRINCE, F. C., MATTOS, T. J. S., and PATEL, N. B. (1982): *Electron. Lett.*, **18**, pp. 1054–1055

QUIST, T. M., REDIKER, R. H., KEYES, R. J., KRAG, W. E., LAX, B., McWHORTER, A. L., and ZEIGER, H. J. (1962): *Appl. Phys. Lett.*, 1, pp. 91–92

RAZEGHI, M., and DUCHEMIN, J. P. (1983): 4th Int. Conference on Integrated Optics and Optical Fibre Communication, Tokyo, Paper 27B4-1

RAZEGHI, M., HIRTZ, J. P., ZIEMELIS, U. O., DELANDE, C., ETIENNE, B., and VOOS, M. (1983): *Appl. Phys. Lett.*, 43, pp. 585–587

RAZEGHI, M., DECREMOUX, B., and DUCHEMIN, J. P. (1984): *J. Cryst. Growth*, 68, pp. 389–397

RENNER, D., GREENE, P. D., COLLAR, A., MOULE, D., and THOMPSON, G. H. B. (1984): 9th IEEE Int. Semiconductor Laser Conference, Rio de Janeiro, pp. 50–51

RENZ, H., WEIDLEIN, J., BENZ, K., and PILKUHN, M. (1979): *Electron. Lett.*, 16, pp. 228

RITCHIE, S., and STEVENSON, A. G. (1984): AGARD Conference Proceedings no. 362, 'Digital Optical Circuit Technology', paper 11

ROSIEWICZ, A., BUTLER, B. R., and HINTON, R. E. P. (1985): *IEEE Proc. Pt. J*, 132, pp. 97–100

RUPPRECHT, H., WOODALL, J. M., and PETTIT, G. D. (1967): *Appl. Phys. Lett.*, 11, pp. 81–83

SAITOH, T., and MUKAI, T. (1986): 10th IEEE Int. Semiconductor Laser Conference, Kanazawa, Japan

SAKAI, K., TANAKA, F., NODA, Y., MATSUSHIMA, Y., AKIBA, S., and YAMAMOTO, T. (1981): *IEEE J. Quantum Electron.*, QE-17, pp. 1245–1250

SAKANO, S., VALSTER, A., TSUJI, S., NAKAMURA, H., MATSUMURA, H., LEE, T. P., and BERGH, A. A. (1986): 10th IEEE Int. Semiconductor Laser Conference, Kanazawa, Japan

SAKUMA, I., FURUSE, T., MATSUMOTO, Y., MATSUMOTO, S., UENO, M., KAWANO, H., and SAITO, F. (1980): 7th IEEE Int. Semiconductor Laser Conference, Brighton, UK, Paper 33

SANADA, T., YAMAKOSHI, S., WADA, O., FUJII, T., SAKURAI, T., and SASAKI, M. (1984): *Appl. Phys. Lett.*, 44, pp. 325–327

SARUWATARI, M., and NAWATA, K. (1979): *Appl. Optics*, 18, pp. 1847–1856

SCHAWLOW, A. L., and TOWNES, C. H. (1958): *Phys. Rev.*, 112, pp. 1940–1949

SCHIMPE, R. (1983): *IEEE J. Quantum Electron.*, QE-19, pp. 895–897

SCHIMPE, R. (1983a): *Z. Phys. B-Condensed Matter*, 52, pp. 289–294

SCHIMPE, R., and HARTH, W. (1983): *Electron. Lett.*, 19, pp. 136–137

SCIFRES, D. R., LINDSTROM, C., BURNHAM, R. D., STREIFER, W., and PAOLI, T. L. (1983): *Electron. Lett.*, 19, pp. 169–171

SEKARTEDJO, K., EDA, N., FURUYA, K., SUEMATSU, Y., KOYAMA, F., and TANBUN-EK, T. (1984): *Electron. Lett.*, 20, pp. 80–81

SERMAGE, B., EICHLER, H. J., HERITAGE, J. P., NELSON, R. J., and DUTTA, N. K. (1983): *Appl. Phys. Lett.*, 42, pp. 259–261

SERMAGE, B., CHEMLA, D., SIVCO, D., and CHO, A. Y. (1986): *IEEE J. Quantum Electron.*, QE-22, pp. 774–780

SHARFIN, W. F., and DAGENAIS, M. (1985): *Appl. Phys. Lett.*, 46, pp. 819–821

SHARFIN, W. F., and DAGENAIS, M. (1986): *Appl. Phys. Lett.*, 48, pp. 1510–1512

SIMON, J. C. (1983): *J. Opt. Comm.*, 4, pp. 51–62

SMITH, C., ABRAM, R. A., and BURT, M. G. (1983): *J. Phys. C*, 16, L171–L175

SMITH, C., ABRAM, R. A., and BURT, M. G. (1984): *Electron. Lett.*, 20, pp. 893–894

SMITH, D. W., and MALYON, D. J. (1982): *Electron. Lett.*, 18, pp. 43–45

SMITH, P. W. (1972): *Proc. IEEE*, 60, pp. 422–440

SPANO, P., PIAZOLLA, S., and TAMBURRINI, M. (1983): *IEEE J. Quantum Electron.*, QE-19, pp. 1195–1199

SPEIER, P., SCHOLZ, F., BENZ, K. W., RENZ, H., and WEIDLEN, J. (1983): *Electron. Lett.*, 19, pp. 728–729

STEPHENS, W. E., JOSEPH, T. R., FINDAKLY, T., and CHEN, B. U. (1984): *Electron. Lett.*, **20**, pp. 424–426

STEVENTON, A. G., SPILLETT, R. E., HOBBS, R. E., BURT, M. G., FIDDYMENT, P. J., and COLLINS, J. V. (1981): *IEEE J. Quantum Electron.*, **QE-17**, pp. 602–610

STOLEN, R. H. (1980): *IEEE Proc.*, **68**, pp. 1232–1236

STOLL, H. M. (1979): *IEEE Trans.*, **CAS-26**, pp. 1065–1072

STONE, J., WIESENFELD, J. M., DENTAI, A. G., DAMEN, T. C., DUGUAY, M. A., CHANG, T. Y., and CARIDI, E. A. (1981): *Optics Lett.*, **6**, pp. 534–6

STONEHAM, A. M. (1981): *Rep. Prog. Phys.*, **44**, pp. 1251–1295

STREIFER, W., BURNHAM, R. D., and SCIFRES, D. R. (1975): *IEEE J. Quantum Electron.*, **QE-11**, pp. 154–161

STREIFER, W., SCIFRES, D. R., and BURNHAM, R. D. (1979): *Appl. Optics*, **18**, pp. 3547–3548

STUBKJAER, K., ASADA, M., ARAI, S., and SUEMATSU, Y. (1981): *Jap. J. Appl. Phys.*, **20**, pp. 1499–1505

SU, C. B., SCHLAFER, J., MANNING, J., and OLSHANSKY, R. (1982): *Electron. Lett.*, **18**, pp. 595–596

SU, C. B., SCHLAFER, J., MANNING, J., and OLSHANSKY, R. (1982a): *Electron. Lett.*, **18**, pp. 1108–1110

SUEMATSU, Y., ARAI, S., and KISHINO, K. (1983): *IEEE J. Lightwave Technol.*, **LT-1**, pp. 161–175

SUGIMURA, A. (1981): *IEEE J. Quantum Electron.*, **QE-17**, pp. 627–635

SUGIMURA, A. (1982): *IEEE J. Quantum Electron.*, **QE-18**, pp. 352–363

SUGIMURA, A. (1983): *IEEE J. Quantum Electron.*, **QE-19**, pp. 930–932

SUGIMURA, A. (1983a): *Appl. Phys. Lett.*, **43**, pp. 728–730

SUGIMURA, A. (1983b): *IEEE J. Quantum Electron.*, **QE-19**, pp. 932–941

SUTO, K., and NISHIZAWA, J. (1985): *IEE Proc. Pt. J*, **132**, pp. 81–84

SUZUKI, Y., and OKAMOTO, H. (1983): *J. Electron. Materials*, **12**, pp. 397–411

TADA, K., NAKANO, Y., and USHIROKAWA, A. (1984): *Electron. Lett.*, **20**, pp. 82–84

TAKAHASHI, S., SAITO, H., and IWANE, G. (1980): *Electron. Lett.*, **16**, pp. 922–923

TAKAHEI, K., KUROIWA, K., and IKEGAMI, T. (1983): *Rev. Elect. Commun. Lab.*, **31**, pp. 321–330

TAMARI, N., and SHTRIKMAN, H. (1982): *Electron. Lett.*, **18**, pp. 177–178

TEMKIN, H., MAHAJAN, S., DiGIUSEPPE, M. A., and DENTAI, A. G. (1982): *Appl. Phys. Lett.*, **40**, pp. 562–565

TEMKIN, H., ALAVI, K., WAGNER, W. R., PEARSALL, T. P., and CHO, A. Y. (1983): *Appl. Phys. Lett.*, **42**, pp. 845–847

TENCHIO, G. (1977): *Electron. Lett.*, **13**, pp. 614–616

THOMPSON, G. H. B. (1980): 'Physics of semiconductor laser devices' (Wiley) pp. 374–390

THOMPSON, G. H. B. (1983): *Electron. Lett.*, **19**, pp. 154–155

THOMPSON, G. H. B., and KIRKBY, P. A. (1973): *Electron Lett.*, **9**, pp. 295–296

TOHMORI, Y., SUEMATSU, Y., TSUSHIMA, H., and ARAI, S. (1983): 4th Int. Conference on Integrated Optics and Optical Fiber Communication, Post deadline papers, pp. 22–23

TOMARU, T., KAWACHI, M., YASU, M., MIYA, Y., and EDAHIRO, T. (1981): *Electron. Lett.*, **17**, pp. 731–732

TRACY, J. C., THOMPSON, L. F., HEIDAREICH, R. D., and MERZ, J. L. (1974): *Appl. Optics*, **13**, pp. 1695–1702

TSANG, W. T. (1981): *Appl. Phys. Lett.*, **39**, pp. 134–137

TSANG, W. T. (1981a): *Appl. Phys. Lett.*, **39**, pp. 786–788

TSANG, W. T. (1982): *Appl. Phys. Lett.*, **40**, pp. 217–219

TSANG, W. T. (1984): *Appl. Phys. Lett.*, **44**, pp. 288–290

TSANG, W. T., OLSSON, N. A., and LOGAN, R. A. (1983): *Appl. Phys. Lett.*, **42**, pp. 650–652

TSANG, W. T., OLSSON, N. A., and LOGAN, R. A. (1983a): *Appl. Phys. Lett.*, **42**, pp. 1003–1005

TSUKADA, T. (1974): *J. Appl. Phys.*, **45**, pp. 4899–4906

TSUKADA, T., NAKASHIMA, H., UMEDA, J., NAKAMURA, S., CHINONE, N., ITO, R., and NAKADA, O. (1972): *Appl. Phys. Lett.*, **20**, pp. 344–345

TUCKER, R. S., CHINLON LIN, BURRUS, C. A., BESOMI, P., and NELSON, R. J. (1984): *Electron. Lett.*, **20**, pp. 393–394

TURLEY, S. E. H., HENSHALL, G. D., GREENE, P. D., KNIGHT, V. P., MOULE, D. M., and WHEELER, S. A. (1981): *Electron. Lett.*, **17**, pp. 868–870

TURNER, J. J., CHEN, B., YANG, L., BALLANTYNE, J. M., and TANG, C. L. (1973): *Appl. Phys. Lett.*, **23**, pp. 333–334

UCHIYAMA, S., MORIKI, K., IGA, K., and FURUKAWA, S. (1982): *Jap. J. Appl. Phys.*, **21**, pp. L639-641

UNGER, H. G. (1971): *Arch. Elek. Ubertragung*, **25**, pp. 539–540

UNGER, H. G. (1977): *Arch. Elek. Ubertragung*, **31**, pp. 518–519

UOMI, K., CHINONE, N., OHTOSHI, T., and KAJIMURA, T. (1985): IOOC–ECOC '85, pp. 493–496

UTAKA, K., AKIBA, S., SAKAI, K., and MATSUSHIMA, Y. (1981): *Electron. Lett.*, **17**, pp. 961–963

UTAKA, K., AKIBA, S., SAKAI, K., and MATSUSHIMA, Y. (1984): *Electron. Lett.*, **20**, pp. 1008–1009

VAHALA, K., and YARIV, A. (1983): *Appl. Phys. Lett.*, **43**, pp. 140–142

VAHALA, K., and YARIV, A. (1983a): *IEEE J. Quantum Electron.*, **QE-19**, pp. 1096–1101 (Pt. 1) and pp. 1102–1109 (Pt. 2)

VAHALA, R., CHIU, L. C., MARGALIT, S., and YARIV, A. (1983): *Appl. Phys. Lett.*, **42**, pp. 631–633

VAHALA, K., HARDER, Ch., and YARIV, A. (1983a): *Appl. Phys. Lett.*, **42**, pp. 211–213

VAN GURP, G. J., DeWAARD, P. J., and Du CHATENIER (1984): *Appl. Phys. Lett.*, **45**, pp. 1054–1056

WALKER, S. D., BICKERS, L., and BLANK, L. C. (1984): *Electron. Lett.*, **20**, pp. 717–719

WALPOLE, J. N., LIND, T. A., HSIEH, J. J., and DONELLY, J. P. (1981): *IEEE J. Quantum Electron.*, **QE-17**, pp. 186–192

WATANABE, Y., and NISHIZAWA, J. (1957): Japanese Patent 273217, as cited in Suto and Nishizawa (1985)

WATANABE, Y., and NISHIZAWA, J. (1960): Japanese Patent 762975, as cited in Suto and Nishizawa (1985)

WEBB, R. P. (1986): CLEO '86, San Francisco, California

WELFORD, D., and MOORADIAN, A. (1982): *Appl. Phys. Lett.*, **40**, pp. 865–867

WELFORD, D., and MOORADIAN, A. (1982a): *Appl. Phys. Lett.*, **40**, pp. 560–562

WESTBROOK, L. D. (1986): *IEE Proc.* Pt. J, **133**, pp. 135–142

WESTBROOK, L. D., and NELSON, A. W. (1984): *J. Appl. Phys.*, **56**, pp. 699–704

WESTBROOK, L. D., NELSON, A. W., and HATCH, C. B. (1981): *Electron. Lett.*, **17**, pp. 952–954

WESTBROOK, L. D., NELSON, A. W., and DIX, C. (1982): *Electron. Lett.*, **18**, pp. 863–5

WESTBROOK, L. D., NELSON, A. W., and DIX, C. (1983): *Electron. Lett.*, **19**, pp. 423–424

WESTBROOK, L. D., NELSON, A. W., and FIDDYMENT, P. J. (1983a): *Electron. Lett.*, **19**, pp. 1076–1077

WESTBROOK, L. D., NELSON, A. W., FIDDYMENT, P. J., and COLLINS, J. V. (1984): *Electron. Lett.*, **20**, pp. 957–959

WESTLAKE, H. J., and O'MAHONY, M. J. (1985): *Electron. Lett.*, **21**, pp. 33–35

WESTLAKE, H. J., ADAMS, M. J., and O'MAHONY, M. J. (1986): *Electron. Lett.*, **22**, pp. 541–543

WHITE, I. H., ASPIN, G. J., CARROLL, J. E., and PLUMB, R. G. (1981): *Electron. Lett.*, **17**, pp. 541–543

WILT, D. P., and YARIV, A. (1981): *IEEE Quantum Electron.*, **QE-17**, pp. 1941–1949

WRIGHT, J. V., and NELSON, B. P. (1977): *Electron. Lett.*, **13**, pp. 361–363

WYATT, R., SMITH, D. W., and CAMERON, K. H. (1982): *Electron. Lett.*, **18**, pp. 292–293

YAMADA, M., and SUEMATSU, Y. (1978): *Jap. J. Appl. Phys.*, **18**, Suppl 18–1, pp. 347–354

YAMAGUCHI, M., KATAMURA, M., MITO, I., MURATA, S., and KOBAYASHI, S. (1984): *Electron. Lett.*, **20**, pp. 233–5

YAMAGUCHI, M., EMURA, K., KITAMURA, M., MITO, I., and KOBAYASHI, K. (1985): OFC '85, San Diego, California

YAMAMOTO, Y. (1983): *IEEE J. Quantum Electron.*, **QE-19**, pp. 34–46

YAMAMOTO, Y., SAITO, S., and MUKAI, T. (1983): *IEEE J. Quantum Electron.*, **QE-19**, pp. 47–58

YANASE, T., KATO, Y., MITO, I., YAMAGUCHI, M., NISH, K., KOBAYASHI, K., and LANG, R. (1983): *Electron. Lett.*, **19**, pp. 700–701

YANASE, T., KATO, Y., MITO, I., KOBAYASHI, K., NISHIMOTO, H., USUI, A., and KOBAYASHI, K. (1983a): *Jap. J. Appl. Phys.*, **22**, L415–L416

YEVICK, D., and STREIFER, W. (1983): *Electron. Lett.*, **19**, pp. 1012–1014

YONEZU, H., MATSUMOTO, Y., SHINOHARA, T., SAKUMA, T., SUZUKI, T., KOBAYASHI, K., LANG, R., NANICHI, Y., and HAYASHI, I. (1977): *Jap. J. Appl. Phys.*, **16**, pp. 209–210

YOSHIKUNI, Y., KAWAGUCHI, H., and IKEGAMI, T. (1985): *IEE Proc.* Pt. J, **132**, pp. 20–27

ZEE, B. (1978): *IEEE J. Quantum Electron.*, **QE-14**, pp. 727–736

Index